Headspace

WITHDRAWN

Amber Marks is a barrister, part-time lecturer in law at King's College London and freelance journalist. This is her first book.

Headspace

Sniffer Dogs, Spy Bees and One Woman's
Adventures in the Surveillance Society

Amber Marks

Published by Virgin Books 2009

2 4 6 8 10 9 7 5 3 1

First published in Great Britain in 2008 by
Virgin Books
Random House, 20 Vauxhall Bridge Road,
London SW1V 2SA

www.virginbooks.com
www.rbooks.co.uk

Addresses for companies within The Random House Group Limited can be found at:
www.randomhouse.co.uk/offices.htm

The Random House Group Limited Reg. No. 954009

A CIP catalogue record for this book
is available from the British Library

ISBN 9780753515549

The Random House Group Limited supports The Forest Stewardship Council [FSC],
the leading international forest certification organisation. All our titles that are
printed on Greenpeace approved FSC certified paper carry the FSC logo.
Our paper procurement policy can be found at www.rbooks.co.uk/environment

Mixed Sources
Product group from well-managed
forests and other controlled source
www.fsc.org Cert no. TT-COC-2341
© 1996 Forest Stewardship Council
FSC

Printed and bound in Great Britain by
CPI Bookmarque, Croydon, CR0 4TD

To Crofton

Contents

PART III

'You will excuse a certain abstraction of mind, my dear Watson,' said he. 'Some curious facts have been submitted to me within the last twenty-four hours, and they in turn have given rise to some speculations of a more general character. I have serious thoughts of writing a small monograph upon the uses of dogs in the work of the detective.'

'But surely, Holmes, this has been explored,' said I. 'Bloodhounds – sleuth-hounds –'

'No, no, Watson; that side of the matter is, of course, obvious. But there is another which is far more subtle.'

Arthur Conan Doyle, *The Adventures of the Creeping Man*

PART I

CHAPTER ONE
The Baying of the Hounds

Lear: Thou has seen a farmer's dog bark at a beggar?
Gloucester: Ay, sir.
Lear: And the creature run from the cur? There thou mightst behold the great image of authority: a dog's obey'd in office.
William Shakespeare, *King Lear*

The first time I was sniffed by a dog I was seven. I entered a room full of grown ups and someone's dog headed straight for my crotch, stuck its nose up between my legs and sniffed. It then wrapped its paws around my leg and started to hump me so hard I fell to the floor. It was my first sexual experience. It was un-consensual and took me by surprise. I blamed myself.

Smell has historical associations with sin, which may be why the experience of being sniffed is unnerving. Patrick Süskind sums up its invasiveness in his novel about a freak of nature called Grenouille who hunts down his victims by their scent and murders them to preserve their human essence for his personal perfume collection. The novel begins when Grenouille is a baby. A priest is cradling Grenouille in his arms, when he wrinkles his little nose:

It was establishing his scent! And all at once he felt as if he stank . . . The child seemed to be smelling right through his skin, into his innards. His most tender emotions, his filthiest thoughts lay exposed to that greedy little nose . . .

The first time the police dog penetrated my consciousness was when I was working as a barrister in connection with a liquor licensing application. It was traditional for licensing sessions to begin with an address to the magistrates by a representative of the area's police force on its latest initiatives. I was sitting on the bench waiting for the officer to finish before I could make my application.

'We will be taking police sniffer dogs around pubs and clubs in the area to detect drug users on the premises,' he stated proudly.

The announcement of this initiative surprised and unnerved me. The British Pub would take on a different character with police dogs sniffing around the feet of those relaxing with friends after a night of debauchery. Had dope smokers lost the right to drink? I made a note to research this initiative.

I might have forgotten this incident had further encounters not swiftly followed.

Walking into Fulham Broadway underground station, I saw officers holding dogs on leashes, encouraging them to sniff the crotches of passing commuters. I approached one of the policemen and asked him what was the purpose of this operation.

'I can't say.'

Luckily, I was wearing a suit and I apologised for my curiosity, explaining that I was a lawyer with a professional interest in crime. He looked at me with something approaching interest.

'Well, you know that most crime is caused by drugs?'

'Yes,' I lied and nodded.

Artificial affinity had been achieved.

'Well, these dogs can smell the smallest trace of a drug on a person. A lot of people take drugs and once the dog has picked up a scent of drugs on them, we have the right to search them. If we find drugs on them we can then search their homes and in their homes we usually find all manner of incriminating articles.'

Now I understood. Who needs a warrant when you've got a dog? Restrictions on police powers were being circumvented by a dog's bark. Man's best friend was a Judas. I nodded approvingly at him.

'The Muslims don't like it though,' he added, 'or the Chinese.' The officer didn't seem particularly upset by this. I thanked him and went on my way.

On my way back from court a few weeks later, I was walking through Clapham Junction station subway when I saw a group of officers with a black dog. The dog handler was tall and chubby with a dirty smile. The officers had stopped to chat amongst themselves just past one of the stairwells. A little black girl stood guarding a suitcase for her mother at the bottom of the steps. The dog kept going over to her case but the handler took no interest. I heard one of the nicer-looking officers ask the handler why it kept going over to her. The handler beamed and laughed.

'He can smell she's black can't he?'

Somewhat revolted by this scene I told my mother about it. It didn't faze her at all. 'Oh I know,' she said. 'I have read about it in *The Daily Bulletin*.'

The Daily Bulletin is the local ex-pat paper in Majorca and often contains snippets from Reuters not picked up on elsewhere in the media.

'In fact there is a case going through the courts at the moment in which the police are being sued for training their dogs to go after black people.'

Did different races emit distinct odours? It seemed unlikely to me but perhaps nothing could be dismissed as impossible in this strange new world of olfactory policing. There was something unnerving about the prospect of being sniffed by a police dog but I couldn't put my finger on what. Perhaps it was the deeply entrenched cultural association between sniffing and snooping. Inquisitive people have historically been derided for their nosiness.

Fed up with the unpaid hours of work required to defend suspects properly, I decided to accept a well-paid office job as a government lawyer in the Court of Appeal. Tracy, a solicitor who had regularly instructed me when I'd worked as a barrister, called me one day out of the blue after I'd been working in the office for several months and we arranged to meet for coffee opposite the Royal Courts of Justice.

'So what are you up to?'

'Bit fed up with the day job – too much administration.'

'That's office life.'

'I'm thinking of undertaking some legal research into sniffer dogs.'

'Really, that will be interesting.' She pulled her Gucci sunglasses back over her heavily made up eyes and took a drag on a Benson and Hedges. 'My client's wife had a run in with them only yesterday.' She passed the packet of cigarettes over to me.

'Do tell,' I begged, discarding my slice of cake and resolution to give up smoking and taking one of her Bensons.

'She went in to visit him in Belmarsh prison. She was sat in the waiting room when the guards walked a sniffer dog past the line of waiting visitors. The dog barked at her and another woman who were then taken into another room. They were told to wait there for female officers to arrive who could search them.' She took a bite of quiche between drags. 'Anyway, they were in there waiting for a while before the other woman turned to speak to her. "I'm worried," she said, "I don't have anything on me but I do have my period." "So do I," replied my client's wife. "Isn't it embarrassing?" Interesting huh? Neither of them had drugs on them but both were menstruating and the dog singled them out.'

'That is interesting. It hadn't occurred to me that the dogs weren't reliable. I've been too wound up about their use in the first place. I mean, since when do the police have a mandate to sniff around in the hope of finding something chargeable? And what about privacy? What's more personal than the way I smell?'

'Have you read Ana Funder's *Stasiland*?' she asked.

'No. What is it?'

'It's a book about the Stasi.'

'Which is what?'

'The Stasi was the secret police in East Germany. Its objective was total control of the population and its means was to know everything about everyone. They turned the German Democratic Republic – I think that's what it was called – into a police state. Everyone living there was terrified of being spied on, informed upon or arrested. The police had a file on everyone.' Tracy took another drag on her cigarette and exhaled pensively. 'It was weird the information they collected. They used it to intimidate people. They scared the crap out of one girl by telling her that they knew her little sister wanted to study music at college. I think they implied they could put a stop to it or something. Anyway, I can't remember what there is on sniffing in her book, but there is something.'

I purchased a copy the next day. In the first chapter, Ana Funder visits the Stasi museum, located in the former headquarters of the Stasi in Leipzig.

> The Stasi had developed a quasi-scientific method, 'smell sampling', as a way to find criminals. The theory was that we all have our own identifying odour, which we leave on everything we touch. These smells can be captured and, with the help of trained sniffer dogs, compared to find a match. . .
>
> Mostly, smell samples were collected surreptitiously. The Stasi might sneak into someone's apartment and take a piece of clothing worn close to the skin, often underwear. Alternatively a 'suspect' would be brought in under some pretext for questioning, and the vinyl seat he or she had sat on would be wiped afterward with a cloth. The pieces of stolen clothing, or cloth, would then be placed in a sealed jar. The containers looked like jam bottling jars. A label read: 'Name: Herr [Name]. Time: 1 hour. Object: Worker's Underpants.'
>
> Leipzig Stasi had collected smell samples of the entire political opposition in this part of Saxony. No-one knows who has these scraps of material and old socks now, nor what they might be keeping them for.

If nothing else, this was material for a viable conspiracy theory on the common problem of disappearing socks. I wondered if the UK authorities had been aware of the Stasi work and whether the increasing number of dogs on our streets was somehow related. I also wondered who had the missing Stasi samples.

I did an internet search on the Stasi. Rumours abounded about its former head being hired by the US Homeland Security Department. The Homeland Security Department is the US equivalent of the UK's Home Office. The Department's activities since its establishment shortly after 9/11 have been controversial and it has been criticised for attempting to turn the US population into a network of spies. I had a vague recollection of Wernher von Braun, the Nazi rocket scientist subsequently taken on by the US to develop weapons of war. Perhaps they were recruiting experts in controlling internal populations for the war on terror.

I started talking to everyone I met about sniffer dogs. Almost everyone had their own story. The topic was so absurd it even worked, I thought, as a method for chatting up men. I started to chase a ridiculously handsome civil liberties campaigner called Tom, with whom I had shared an animated pub conversation about police dogs. We met for a drink and weren't able to re-capture the energy of our first conversation. I got very drunk and the date, if that's what it was, ended with him carrying me home while I vomited over his shoulder. I was too embarrassed for a long time after that to contact him again. I wouldn't hear anything from him and then out of the blue my phone would ping and a text from him would read: 'Guardian. Page 5. Good doggy story xxx'. It slowly dawned on me that I had trained this man to think of me every time he saw a dog; not the best pick-up trick in the book.

A barrister friend of mine was representing a man charged under the Cruelty to Animals Act of 1874 for allegedly hitting a sniffer dog on the nose. Proof that the dog had been injured was the dog handler's evidence that the dog had yelped. There was scope for some decent cross-examination there.

A man I met in a club was leaning against the wall of

WHSmith in Victoria station when a dog came and sat next to him. The dog's owner, a policeman, approached him and told him he was lucky his dog was still in training for passive drug detection duties, otherwise he would have had to search him. Passive drug detection, I was later to learn, was the latest fashion in dog training. Instead of barking at its suspect, the dog was trained to sit down next to them.

A *Big Issue* vendor told me his life was destroyed by a sniffer dog. Homeless and fleeing drug addiction in London, he was boarding a train to his parents' house when a sniffer dog barked loudly at him. The police came running over and asked him for his details. They looked him up, found out he had outstanding warrants, hauled him off the train and locked him up. He was unable to fathom the scent that had alerted the dog to him.

A friend and heavy cannabis smoker had been stopped every morning on his way to work by barking dogs and dog handlers asserting a right to search him at Seven Sisters station. On the twelfth occasion he objected, complaining they made him late for work and they should know by now that he never had anything on him. The dogs continued to bark and the officers persisted in searching him on a daily basis. Eventually he contacted a solicitor. The solicitor wrote to the police and the searches stopped.

Friends, aware of my interest, started to send me articles they came across.

One was about a headmaster who had told his pupils to line their bags up outside and congregate in the assembly room. While he talked, police officers, who had been invited on to the school grounds, had their dogs sniff the children's bags for drugs.

I thought I'd phone the Home Office to find the official policy on sniffer dogs.

'Hello. Can you tell me if you have any guidance on sniffer dogs?'

The man on the end of the telephone sniggered.

'On what?'

'Sniffer dogs.'

'No, I don't think so.'

'Well, do you have any publications on sniffer dogs at all?'

'No, I don't think so,' he laughed again.

'Well that is odd, because a large number of police forces seem to have started using them. Are you sure the Home Office doesn't have any information on them?'

'Not that I have come across. I suggest you try the police.'

I phoned up a friend and former head of the drug squad.

'What do you know about sniffer dogs?'

'Nothing, never had anything to do with them.'

'But it must have been when you were in charge that the police started to use them for finding drugs?'

'As I say, I don't remember having anything to do with them. A friend of mine runs a dog training school though. I'll have a word. I'm sure he won't mind you making a visit.'

'Fantastic,' I said. 'Thanks very much.'

Over Christmas in Majorca I had coffee with an ex-boyfriend who now works as a policeman. I told him about my research. He told me that they were setting up a canine unit on the island. He was sure it was for explosive detection only. I had my doubts given the ever-increasing demands put upon law enforcement by the War on Drugs.

I met up with a local legalise cannabis activist, and tried to alarm him with stories of whom the dogs were being used against in the UK and the news that a canine unit was planned for Majorca.

'*Tranquila Ambar.* They won't do that here. Not in

Majorca. I tell you why. We had a protest recently. It was stopped by the police who beat the protestors with sticks. The next day, one of these policemen goes round to his uncle's house and sits down at the kitchen table. "Get out," shouts his uncle. "You can't beat me with a stick one day and expect a meal the next." Majorcans would never allow the state to sniff our balls. If they start trying to behave here like that with those dogs, the dogs will be killed. "Who killed my dog?" the officer will ask. "What dog?" they will reply.' He shook his head and walked across the room. 'No, the Majorcans won't stand for it, even less so if the dog is not a Majorcan breed.'

I contacted the civil liberties organisation where Tom worked and got put through to him. I hadn't talked to him since a brief conversation the day after our disastrous date.

'Hello, it's Amber.'

'Hello, how are you?'

'Good. Good. Much better thanks. Still very embarrassed and grateful.'

'Honestly, Amber, it happens to everyone. It was no trouble at all. How have you been?'

'Busy, busy. You know how it is.' I haven't been thinking of you day in day out. I am not phoning you on the pretext of work just to talk to you. This dog-sniffing thing is serious. 'I was wondering if you had carried out any research into the civil liberties implications of the increasing use of sniffer dogs.'

He laughed. 'Animal rights you mean?'

As far as I was aware, his organisation didn't campaign for the rights of animals.

'No. I think the use of police dogs in an increasing range of public settings has implications for all of us.'

'Well, you can write something on it for our website if you want. We could probably fit it in during our silly season. No offence but it does have comedy value.'

'OK, I'll be in touch.'

Was I wrong to take this dog thing seriously? Was my obsession with dogs a sophisticated twist on my crush on this man? Was I barking up the wrong tree?

Then the Abu Ghraib story broke. The papers and television news were filled with photos of Arab prisoners being tortured by American soldiers. Journalists ranted about the hoods, knickers and nudity. I noticed the large Alsatian dogs in the photos, and remembered what the officer had told me in Fulham Broadway station about Muslims not liking them. Historians researching the mythology of the dog in Anglo-Saxon, Viking and medieval times claimed that its use by marauding invaders had cemented it in the popular psyche as a 'potent embodiment' of the threat to an existing social order. An article in the *Observer* described the police dog as 'the public face of the War on Terror'. I was sure there were dark forces afoot in this dog business and I decided it was time for some serious research.

Swarm of bees grounds UK aircraft

BBC NEWS
25 May 2007

Almost 200 passengers found themselves stranded at Bournemouth Airport for 11 hours after their plane turned back after flying into a swarm of bees.

CHAPTER TWO

The Psychic Connection

I can feel that least black boy out there coming up the hall, smelling out for my fear. He opens out his nostrils like black funnels, his outsized head bobbing this way and that as he sniffs, and he sucks in fear from all over the ward. He's smelling me now. I can hear him snort. He don't even know where I'm hid, but he's smelling and he's hunting around. I try to keep still.

Ken Kesey, *One Flew Over the Cuckoo's Nest*

Several months later, frustrated with my day-job and increasingly fascinated by the historical use of working dogs, I found myself cycling home behind a bus with a British Transport Police poster on the back of it announcing the advent of the Year of the Dog. The poster showed a police officer with a dog on a leash and wished me a Happy New Year. The next day I phoned up the press office of British Transport Police and asked where I could find the policy initiative behind the increasing deployment of drug detection dogs in London stations. They told me there was none and that the increase was simply due to the fact that they had more dogs these days.

I was wary of this growing army of police dogs, having discovered a long-standing relationship between dogs and authoritarianism. Christopher Columbus had introduced dogs to North America to terrorise the natives. Further use was made of them during the American Civil War when they were let loose on runaway slaves. In Nazi Germany the SS conditioned their dogs to savage anyone who was not in uniform. Himmler chose to arm guards in female concentration camps with dogs instead of guns on the basis that dogs would frighten women more. He ordered that the dogs be used to suppress the first sign of revolt. Forced sex with trained dogs was used in Chile, under Pinochet's rule, as a form of torture. In 1950s North America, rock 'n' roll concert attendees were greeted by snarling vicious dogs in the guise of 'agents of society's disapproval' and in the 1960s dogs were routinely used on protesters against the Vietnam War and during race riots.

The US Defense Department has observed that the mere presence of a dog can create anxiety and that dogs can be used as both a psychological and physical deterrent. According to US officials, military working dogs are 'effective in setting the atmosphere' for obtaining information. Dogs were, and may still be, used to induce fear and stress in detainees at Abu Ghraib prison in Iraq. A company currently providing services to armies around the world claims to make handler and dog function as a single entity through its employment of 'psychic and mental methods from the Eastern Bloc'. Its dogs are imported from the Czech Republic and Slovakia and the company claims that they contain a 'switch' to go 'from nice to nightmare'. The dogs are described as 'the best instrument' for the 'War on Terror'. Was this because of their effect on the atmosphere? Given the repressive connotations of the dog, I found it surprising that the police in democratic countries are so keen on using them in schools and stations. I hadn't appreciated that this was where the War on Terror would be waged.

I attributed the lack of controversy over the use of dogs to the fantastic PR job done on them. Potentially lethal, they were also man's best friend. According to an article in *Fortune*, an American business magazine, there was just something about a dog's word: 'From Lassie to Rin Tin Tin, all the way back to Odysseus's dog Argos (the only one to recognise Homer's hero after his 20 year absence), our culture is littered with testaments to the honest dog and his reliable nose.' Judging from an early twentieth-century account, cited in Martin Cotter's *All About German Shepherd Dogs*, the Dog Detective was an early embodiment of Poirot:

After taking a few deep sniffs around the blood-stained bed the dog set off down the stairs to the yard below, where all the staff, male and female, were assembled and were standing apprehensively, in a close group. No one dared to move, as the dog approached and started circling the group, snuffling deeply. The tension was reaching breaking point when the dog started to growl and focused his attention on a particular person. Growling menacingly the dog moved closer to the man, while the rest held their breath, petrified. Then the man's nerve snapped and he made a furtive movement. Instantly the dog sprang and gripping him by the thigh, brought him to the ground. The dog had made no mistake; in addition to the first scent picked up by the bedside and tracked to the man, fear and guilt had made the adrenaline flow in a peculiar mixture and the chemical reaction gave off an odour which the dog recognized. The dog understood the furtive movement entitled him to make an arrest and not let the man get away.

Pleading for the dog to be called off, the murderer made a complete confession of his crime, before everyone right there on the spot. He was sentenced to death and while awaiting execution, told his gaolers he did not fear the gallows nearly so much as he did fear the dog, with flashing teeth and snapping jaws, which, ever and always haunted him in his sleep.

It sounded like he had confessed out of fear to me and I wondered if he could have been innocent.

I'd heard the phrase 'Pavlov's dogs' and knew it was important stuff, but I hadn't had cause to look into it before reading that current dog training is still based on Pavlov's early twentieth-century methods of Classical Conditioning. Pavlov was a Russian psychologist and physiologist whose main interest was the control of the nervous system. He observed that dogs salivate when they eat. This he described as a simple reflex action. He further observed that dogs would salivate at the sight of food. He hypothesised that this was a more complicated response than a mere reflex action and that it involved a psychical event. The dog, he argued, had learned by experience to associate that sort of visual stimulus with eating. It occurred to Pavlov to experiment with whether this one-link chain of association could be lengthened. Thus he would start by ringing a bell before presenting food to the dog. After a sufficient number of such experiences the dog would associate the ringing of the bell with the appearance of food, and would salivate when the bell rang and before the food had appeared. Such a reaction is called a 'conditioned reflex'. Pavlov tried various stimuli – noises, colours, shapes, touches on various parts of the body, electric shocks – and found that the dog could learn to regard any such stimulus as the prelude to being fed. Ironically, given the current applications of Pavlovian trained dogs, the only stimuli he found difficult to apply with any reliably predictive value were smells:

> It has been exceedingly difficult, if not impossible up to the present, to obtain the same accuracy in graduation of olfactory stimuli as of any other stimuli. It is impossible also to limit the action of olfactory stimuli to any exact length of time. Furthermore, we do not know of any subjective or objective criterion by which small variations in intensity of odours can be determined . . .

In the course of his experiments, Pavlov observed that dogs exhibit all the symptoms of a nervous breakdown when subjected to prolonged physical or psychic stress. Shortly before reaching this point of final breakdown, they become more than normally suggestible. This is when new behaviour patterns can be installed most efficiently. Those installed at this juncture will be ineradicable; that which the dog has learned under stress will remain an integral part of its make-up. Pavlov's speculation that the actual cells of the central nervous system might change structurally and chemically when a new behaviour pattern is formed has since been confirmed by neuroscientists. This would no doubt account for the anecdotes I'd heard about ex-sniffer dogs continuing to seek out cocaine in their retirement. Their training had become hard-wired. Even more interesting was Pavlov's conviction that trained dogs could pass their lessons on through their genes, endowing future generations with freshly conditioned neural links. This theory was adopted from Jean-Baptise Lamarck, a nineteenth-century zoologist who coined the term 'biology' and divided the animal kingdom into verte-brates and invertebrates. Lamarck's theory of evolution, which even Darwin found persuasive, is known as the 'inheritance of acquired characteristics'.

So immersed was I in my research that I'd almost forgotten Tom. Then he texted me. He'd had an encounter with a sniffer dog. Serves him right for not taking the topic seriously, I thought, but I couldn't get the possibility of speaking to him out of my mind. I made my way outside to the foyer of the British Library, sat on a cold marble plinth and dialled his number.

'Hey Tom, it's Amber.'

'Amber. Hey, how are you?' Strangely, he sounded surprised to hear from me.

'Good thanks. Been busy. I gave a talk to lawyers on the legality of sniffer dogs the other day. It actually raises all sorts of interesting legal issues.' I felt like I was babbling defensively to prove I hadn't been thinking about him pointlessly all day every day since we'd last spoken. 'How are you?'

'All right. Sorry I haven't been in touch for ages. I've been working non-stop. I'm orchestrating a campaign against Total Surveillance. You wouldn't believe how quickly we are becoming a police state. One company is even running a fingerprinting system in British schools. Guess what the system is called.'

'I don't know,' I replied pathetically. I felt the loss of all capacity for dialogue whenever I spoke to him and it made me nervous.

'It's called VeriCool.'

'No way!'

'Yeah.'

'How do schools justify taking their kids' fingerprints to the parents?'

'All sorts of bizarre ways. One school claimed it was a simple and fun way for the library to keep track of who had what books.'

'That's absurd.'

'I know. Oh by the way, I found out where the expression barking up the wrong tree comes from.'

'Where?' I laughed, thinking that perhaps he did love me after all.

'Apparently the doggy has been getting the wrong end of the stick for donkeys' years.'

'What do you mean?' I asked.

'It's an early nineteenth-century expression from when they used dogs to hunt racoons. The dogs were trained to bark at the foot of whichever tree the racoon had run up. It's a reference to the waste of the huntsmen's time and energy by the dog barking up the wrong one.'

'Well you know,' I said excitedly, 'they do seem to be wrong a lot of the time. I got talking to someone from Action on Rights for Children recently as they've had a number of complaints from innocent children searched as a result of police dog indications. Apparently one got told, after he was searched, that he must have come from a venue where drugs were being used. He was able to prove he'd spent the day at the Department of Education and the police officer finally apologised for his accusatory manner. But I interviewed the press contact for the British Transport Police the other day and asked him how often the dogs got it wrong and he said it was "impossible to tell". He said that when the person is found not to have any drugs on them it is because the dog is "so sensitive" it can detect the scent on you if you have brushed past someone who is carrying drugs. So even when they are wrong they are right.'

Tom explained that a number of detection technologies fell into the 'over-sensitive' category. They enabled the police to investigate and ask innocent civilians to explain their whereabouts and provide personal details by triggering false positives. People were normally so alarmed by having triggered a response from the detection device that they divulged all manner of personal information. The information was then entered on to a record and stored for future use. He said it was very similar to the Total Information Awareness programme in the US.

'Sounds like the real reason for the large-scale deployment of dogs is to widen the net of surveillance over the general population. Pigs using dogs to fish humans!' he joked.

Finally, I thought. He was beginning to see some relevance in my research.

'So what encounter did you have with a dog?' I asked.

'I was coming out of Highbury and Islington station and saw police with dogs to the left and so went right. The police followed me and stopped and searched me.'

'What reason did they give for the search?'

'They said they saw me change my direction on spotting the dog.'

'That's not a valid reason for stop and search. Any chance I could have a copy of your search record?'

'Absolutely. It's yours. We should meet up anyway.'

'Yeah. That would nice.'

'Maybe get some dinner or something.'

'Sure,' I said, hoping he'd suggest a date.

'OK. Speak to you soon.'

'OK. Bye.'

He hung up. I'd forgotten to tell him that dogs were historically symbolic of authoritarian regimes. I couldn't call him back just to tell him that. He'd think I was nuts. Total Surveillance? What sort of information would be useful to an authoritarian regime?

My thoughts were redirected by the sound of my telephone ringing. It was a withheld number.

'Hello?'

'Is that Amber Marks?'

'Yes.'

'Hello. This is Clare from the mental health trust. A colleague gave me your details. He heard a talk you gave on sniffer dogs at a seminar. I understand that you know quite a lot about them.'

'Yes.'

'Well then it probably won't surprise you to hear that there is a lot of enthusiasm for using them in psychiatric wards. I have to draft our trust policy on their use and I was hoping you would be willing to help me.'

Knowing what I did about dogs, I was amazed that the National Health Service was planning on using them on such a vulnerable section of the population.

'Why is there so much enthusiasm?' I asked. 'Aren't you concerned about the atmosphere they'll create on the ward?'

'HM Prison Service has been using mental health hospitals as training grounds for their dogs for some time now.'

This struck me as odd.

'We are looking,' she continued, 'for measures to keep drugs out of our wards and so we are thinking of introducing dogs.'

'And why do you think dogs are an effective means of keeping drugs out?'

'Initially I was against them. I mean, they are used in prisons and prisons remain full of drugs. But after listening to our Criminal Justice Advisor I can see why they are a good idea. We could get ourselves into legal difficulties unless we can show we are taking all reasonable measures to keep illegal drugs out of our premises.'

'Who is your Criminal Justice Advisor?'

'Walter Penay. He's a retired police officer and is very knowledgeable about criminal justice matters.'

In my experience, police officers are rarely that clued

up on the law and the rights of mental health patients is a complicated area. No doubt he was very much aware of the illegality of drug use, but what about human rights? Clare gave me Walter's phone number and I left a message asking if he would be willing to meet and share his expertise on dogs with me.

The next morning I picked up a voicemail message from Walter. He was free to meet at lunchtime. A few hours later we were sitting in a cafe, sipping apple juice. He informed me that his consultation had concluded and he had recommended the use of drug detection dogs on wards on a pilot basis.

'I've got the police to agree to provide them for free.' The possibility of the police charging for this service hadn't even occurred to me.

'What evidence did you find that they were an accurate detection tool?' I asked him.

'None,' he said firmly. 'But from my years of experience in the police force and working with units using dogs I know that they are very good. Handlers would put money on their accuracy. And,' he added, 'the dogs act as a deterrent to drug use. They are much more efficient at detection than police officers and everyone knows that they are. They may be attributed skills that they don't have but so what?' He shrugged contentedly.

'How did the patients you consulted feel about the idea?'

'Some complained that we were treating them like prisoners. Some thought it might heighten the paranoia of patients on the ward.'

'Isn't that a serious concern?' I asked.

'There is a natural fear of dogs in some people,' he explained. 'It is like the fear of spiders.'

'So the patients don't want dog visits?' I wanted to change the subject from spiders.

'When we *first* mentioned the idea of bringing dogs in a lot of people had images of rottweilers and German shepherds. In fact the dog used is a springer spaniel, which is very small, lovely and fluffy.'

I'd read up on the springer spaniel, having spotted them with police on the underground. It was a 'reliable hunter', a 'family friend'; or, in the words of the poet Jon Gay, a 'fawning slave to man's deceit'. Evidently it was now taking on a new role, as an institutional pet; a 'noser out of unorthodoxy', to borrow a phrase from George Orwell.

'In fact,' Walter continued chirpily, 'taking the dogs in is usually enjoyed by the patients. They usually want to pat the spaniel. It is good for the relationship between the patients and police officers because the patients see that police officers are human, often chatty and don't have two heads!' he explained.

A police officer accompanied by a dog seemed an odd way to convince persons of unsound mind that police officers were human and didn't have two heads but I wasn't there to argue. Busting mental health patients for using drugs of their choice also seemed an odd way to foster good relations between them and the police.

'I've got a picture of one of our spaniels here,' he said, reaching into his black leather briefcase and taking out a poster. 'We put these posters up around the wards months before taking the dogs in.'

A curly-haired spaniel was pictured on the grass. Its eyes were red. A lead was attached to its collar but the picture did not include to whom or what it was attached. 'Drug Dogs are Used on These Premises', read the caption.

'What is the purpose of these posters?'

'It tends to have the effect of fostering acquiescence to their use. And it tells them what the dogs are looking for.'

'I see.'

'Of course, there is a possibility that the dog picks up on the scent of fear but this is something you would have to ask a handler, you know, get from the horse's mouth, or the dog's mouth rather!' he joked.

Why had dog handlers been using mental health wards as training grounds? Why did he think that the dogs were picking up on the scent of fear? I remembered my mother, who had spent two years in prison when I was a child, telling me she could now sense when she was in the vicinity of a prison because she recognised the smell of fear. Was there such a scent? I thought about *Nineteen Eighty-Four*, Winston Smith, and his fear of rats. How had Big Brother known Winston was scared of rats? Had the rats told him?

All Shook Up: Guard dog mauls Elvis's teddy in rampage

Guardian
Maev Kenedy
3 August 2006

When Barney met Mabel, there was an instant – and fatal – chemical reaction.

On Tuesday night the Doberman Pinscher guard dog, after six years' blameless service, went berserk: within minutes Mabel, a 1909 German-made Steiff teddy bear once owned by Elvis Presley, more recently the pride and joy of an English aristocrat, lay mortally wounded.

Barney went on a rampage through dozens of rare teddies, all on loan to Wookey Hole Caves in Somerset, and so valuable that the insurers had insisted on a guard dog to protect the premises at night. 'It was a dreadful scene.'

Barney's mortified handler, Greg West, who took 10 minutes to get the dog back under control, said: 'I still can't believe what happened. Either there was a rogue scent of some kind on Mabel which switched on Barney's deepest instincts, or it could have been jealousy: I was just stroking Mabel and was saying what a nice little bear she was.'

CHAPTER THREE

Best in Show

'He was running, Watson – running desperately, running for his life, running until he burst his heart and fell dead upon his face.'

'Running from what?'

'There lies our problem.'

Arthur Conan Doyle, *The Hound of the Baskervilles*

The man put his newspaper down and smiled at me through the front passenger window. He was in his late fifties and balding, his round eyes framed in black plastic. I was squatting on the pavement, my eyes level with his, cigarette in one hand and my cotton dress held up against the back of my thighs in the other.

'Could you take me to Newbold Revel?'

'I certainly can.'

'I've just lit this cigarette. Do you mind waiting?'

'Get in,' he reached over to open the car door. 'You can smoke it in here.'

'Thanks,' I said. I was still grappling with the frustration left over from last night's dinner date with Tom. If it was a date – we'd gone for pizza.

'Could we stop at a bank on the way?'

'We can rob a bank on the way!'

'Great!' I said, getting in and closing the door. 'I've come to Rugby for an adventure.'

'What are you going to the Prison Training School for? You do realise that today is Saturday?'

'Yes, I do. I'm going to the Annual Police Service Dog Competition.'

'But you haven't got a dog.'

'I know,' I said, disappointment leaking out of my voice.

'Do you want to borrow mine?'

'I'd love to borrow yours. Do you really have one?'

'I have got a fine dog,' he said, reaching over the gear stick and slapping my thigh. 'The police came round for him a few months back. Wanted me to volunteer him for a police trial. Detective friend of mine has had his eye on him since he was a pup. He is so fierce looking he'll scare the shit out of anyone that sees him. If you so much as point a finger at me he'll growl at you. When my wife starts on me I look at Rollo and he starts growling at her. I wouldn't hand him over to the police for a million pounds.'

'He sounds perfect.'

Unfortunately he was joking about lending him to me. I hadn't socialised with dog handlers before and a dog might have provided us with some common ground. The research I had done into dog handlers suggested that we might not get on. In fact, the more I read the more I was convinced that the police dog handler must be a peculiarly twisted form of animal lover. Prospective police dog handlers are screened for liking dogs. Most police dogs live with their police officer handlers, sometimes sharing a bedroom. Despite this, the relationship between the dog and his handler is viewed as a professional one and dogs are referred to as 'colleagues' in police statements. Most animal lovers will tell you at some point how animals are 'better than humans', but dog handlers take matters a step further. They strive for medals in competitions where the criterion for success is the number of humans locked up.

One dog handler has been reported as having admitted to masturbating his dog as a reward and I noted that under the most recent Sexual Offences legislation, animal masturbation is not an offence. The term 'handler' was developing disturbing connotations in my mind.

Many police dog handlers claim their dogs have 'magic' powers. This belief in their dogs' infallibility appears to be genuine. In 2003, Russell Lee Ebersole was convicted of fraud for supplying the US government with useless sniffer dogs. Mr Ebersole had boasted that he had revolutionised dog training by creating animals so highly skilled they could tell their handlers which drug they had detected by pointing to plastic letters of the alphabet with their noses. The dogs were duly purchased and the fraud went undetected until an anonymous tip-off was received. The handlers must have convinced themselves that the dogs were making references to street slang terms for drugs when the dog pointed at letters that didn't make immediate sense. I thought about what Walter Penay had said about handlers being willing to put money on the accuracy of their dogs. The system clearly wasn't foolproof.

It wasn't hard to penetrate the social scene of the dog handlers. Most of them subscribe to a police and services dog club monthly magazine. Membership of the club is restricted to members of the law enforcement community but it is possible to become an Associate Member if nominated by two members. I emailed the editor of the magazine, telling him how much I enjoyed reading it and expressing my desire to become a member so that I could have it delivered to my door. He had a word with the membership secretary and in due course a lapel badge, gold membership card and invite to their annual police dog trials arrived in the post.

The training grounds were a twenty-minute drive from the station during which my driver told me about his mobile disco. He had a smoke machine, a karaoke machine and over 8,000 CDs.

'As soon as a new song comes out, I get it,' he boasted, pulling up alongside the field where the dog trials had evidently already started. He pointed to one of the dogs. 'My dog is five times the size of that one. Rubbish dog that is.'

'The German shepherd?'

'That's no German shepherd! That's an alsatian. Look at its ears. That's how you can tell.' I had no idea what he was talking about.

Some thirty spectators were in attendance. I had expected a larger, more high-profile affair, having seen the size of the trophies from photos of the winners in national newspapers and the members' magazine. This looked like a vicar's garden fête. Tea and orange squash were being served in the refreshment tent. The sky was overcast and the temperature had dropped some ten degrees since I'd carefully picked out the outfit I'd guessed to be most appropriate for this sort of event. As it turned out, I was the only person in a dress. The few women there were wearing either jeans or baggy beige trousers. Those serving tea cooed at the men's military, prison guard and police uniforms. Among them, one man stood out. Almost seven feet tall, he was dressed in a dusty blue boiler suit.

I made my way over to the club members' stall. The people manning the store were busy with customers purchasing teacups with the club's logo on them. A man in a cap of similar design was standing at the side of the stall and smiled at me.

'I hadn't realised how cold it would be today,' I explained. 'I wonder how much those jumpers are?'

'John! This young lady wants to know how much the jumpers are!'

'£21.50.'

'I don't suppose you take cards do you?'

'No, come back later if you want though. I'll save you one.'

'The lady is cold, John.'

'Oh, right,' he said, looking over at me. 'Are you a member?'

'Yes.'

'What is your name?'

'Amber Marks.'

'Aha. You made it. Dave!' He called over to a chubby smiling photographer and pointed at me. 'It's the lawyer!'

Dave smiled and we waved at each other.

'Have you got your number?' John asked, pen in hand.

I pulled out my membership card.

'Very organised,' he said, writing down my number and handing me a bright red jumper. In the top left-hand corner was a picture of a German shepherd dog, the name of the club emblazoned in a circle around its fine physique. I quickly pulled it over my dress.

'That looks great on you!' beamed John. 'Can we have a photo of you in the courts in that for the magazine?'

'Sure.' I loved the idea. 'In my wig as well?'

'Cor, she is serious,' giggled the man who'd persuaded John to give me the jumper on credit.

'Right,' said John. 'Make the cheque to the club and post it to the address in the front of the magazine.'

'Will do. Thank you.'

'All right. Don't forget. I've got three large German shepherds.'

I laughed nervously and made my way to where the spectators were standing. They were watching uniformed police officers take the dogs round an agility course. They had to run up ramps, walk over seesaws, run through tunnels and weave in and out of bollards. I was surprised by how disobedient a lot of them were. A lot of the dogs wouldn't even go up the ramp. I'd heard these were the best police dogs in the country.

I got myself a glass of orange squash from the refreshments tent and sauntered around, hovering on the outskirts of the various social groupings.

'. . . The drug detection market is yet to be tapped. Sure, some people have done the odd thing, Grosvenor International Services for example, but I think there is a lot more potential to be tapped. I mean drugs in schools, people get a bit touchy feely about schools cause they are little and that but at the end of the day you either want the school clean, or you don't.'

'Absolutely.'

'The thing to do is to send letters to the parents stating "if you want your kids' human rights to be protected, if you want them to be free to learn and get a good education, they need to be in a drug-free environment." Ask them to indicate whether they consent to drug dogs being used at the school. When you have got the forms back, if seventy per cent say yes . . .'

'That is a majority.'

'Exactly right. A lot of the time it is the kids who want the dogs. The difficult issue is whether to involve the police when drugs are found. I don't think there is a need. I think that is something schools need to realise . . .'

I took a quick look at who was speaking. A short fat

man in a baseball cap was doing most of the talking. His T-shirt bore the name of a private security dog firm. Two taller grey-haired and otherwise nondescript men were nodding their heads.

'Well, apparently,' said one of them, 'insurance companies are delighted at the increasing number of fire service dogs. Saved them fortunes.'

I walked back over towards the club's stall and asked John where the scent trials were taking place. He directed me to the other side of the car park.

The car park was chock-a-block with barking vehicles. The back doors to some of the vans were open and caged dogs snapped and growled at me as I passed them. A group of uniformed policemen were chatting at the edge of a patch of field. The drug trials were over and the amphetamines were being swapped for an accelerant in preparation for the next competition. I hadn't heard of accelerant sniffer dogs and didn't know what they did. I waited with the officers as the pots were prepared. There were three rows, each containing eight chambers. I walked over to the judge, who was carefully taking items out of plastic bags and placing them inside the chambers.

'Excuse me,' I said.

He looked up at me.

'Yes?'

'What are you putting in the pots?'

'Well, I am putting an accelerant in this one. I am using white spirit. And distractor scents in the others.'

'What are distractor scents?' I asked.

'Oh, different things,' he rattled off a list of arbitrary items, 'and sausages.'

A murmur went up in the group of policemen. I went back over to them. They were talking about the use of sausages.

'I don't think my dog would be able to succeed in the

competition. Sausages never featured as a distractor scent in his training.'

'No. It's a tough one.'

Their tone was serious. They were relieved that they were spectators rather than competitors today.

A tall man in a tight-fitting fireman's outfit arrived with a Labrador on a lead and gave his name to the judge who noted it on his clipboard. The fat man whom I'd overheard talking about the gap in the drug market turned up.

'Is he the only competitor in this trial?' he asked the judge who nodded in response. 'Shall I enter one of mine then?'

The judge agreed and noted his name, Derick Paley, on the clipboard. The fireman went first. He walked his dog confidently up to the chambers. His dog didn't take long to indicate at the correct one. The company man's dog was hopeless. It toppled the chambers over and failed to find the accelerant.

I approached the fireman and asked him what accelerant dogs were used for. Apparently they are the first step in a fire investigation. They can find an accelerant in the rubble in minutes. If white spirit is found in the middle of the sitting room that would suggest the fire was deliberate. No wonder insurance companies were in favour of them.

'I've got a bloodhound now though,' he volunteered.

'Oh have you?' I didn't know why he had mentioned this. 'What do you use it for?'

'Human scent,' he smiled proudly.

'How will that be used in the fire services?'

'No, it's a sideline.'

'You will use it for private work you mean?'

'Yes. For human tracking. Some legitimate,' he paused and smiled conspiratorially at me, 'some not.'

I didn't know what he meant and it felt inappropriate

to ask. He'd said it as though I should know what he was talking about. The only illegitimate tracking I could think of was that described in Arthur Conan Doyle's *Hound of the Baskervilles*. Surely he didn't train his dog to track down individuals disliked by his private clients?

The myth behind the Hound of Baskerville's deadly attacks was that it was a phantom beast that sought out descendants of Hugo Baskerville to punish them for his sins. In fact, as Sherlock Holmes eventually deduced, a distant relative of Hugo Baskerville, masquerading as an unrelated neighbour, coveted the ancestral land. In order to get rid of those who stood before him in the line of succession, he trained a huge and terrifying dog on their scent and released it on the moors. Fear of the hound's legend compounded its attacks and led to the death of its victims. The observation that psychological stress can increase mortality through heart attacks has since been given the term 'the Baskerville effect'.

Absorbed in thought and lost for words, I thanked him for the chat and strolled back over to the stalls. A man in a deckchair was reading a magazine entitled *Protecting World Citizens*. The front cover was emblazoned with a photo of a German shepherd dog baring its fangs and dribbling saliva from its jaws.

'Excuse me,' I asked him. 'Where did you get that magazine?'

He winced up at me. His face was weathered and he looked displeased at my interruption.

'It is a secret magazine. There are more in the box under that table.' He pointed his arm towards the club's stall. I wandered over, asked John if he had a secret magazine under his table and was met with bafflement. There were no magazines in their boxes, other than the back issues of their own publication, with which I would already be familiar.

I strode back over to the man in the deckchair. A horse-faced woman in her early fifties was standing beside him and was now holding the magazine. She had short wavy, mousy coloured hair and wore navy trousers and a khaki combat vest.

'It looks like you have the only copy of that magazine,' I said to the man in the deckchair.

'Hmmn,' he smiled.

'I'd really appreciate a look at it at some point. The only security dog magazine I have come across apart from our own is the American K9.'

'Ah, we were just talking about that,' interjected the woman with bristled enthusiasm. 'Bit OTT the K9.' She didn't look up from the magazine when she spoke.

'So is that one,' said her companion, getting out of his deckchair. 'I was surprised to find out it was English. It's got an article in it on how to kill a dog.'

'Has it really?' tittered the woman. 'Where is that then, Clive?'

Clive found the article before turning his back on her to face me.

'Yes. It says you have to wrap the dog between your thighs and squeeze as hard as you can.'

The woman either didn't catch what he said, or failed to react to his comment.

'Really? That sounds difficult.' I frowned naïvely. 'How would you have time to get your legs into position when the dog is running at you?'

'With a kung-fu move. Do you want to try it during the man work trials?' Clive flashed his teeth in a dirty smile.

'I think I'd need to practise on some smaller dogs first.'

He looked at me good-naturedly. 'I'll lend you the magazine but please put it back in this box when you have finished.'

'He has a large German shepherd that will track you down if you don't.' The woman wagged her finger at me and handed me the magazine.

'I'll have read the article by then and will know how to kill it.'

They both laughed and made their way to where the man work trials were starting. I sat down in the man's deckchair and read the article on killing a dog. It made no mention of thighs. It didn't tell you how to kill it, it simply advised you not to hesitate before doing so if the dog was about to attack you.

'Oi!' someone hollered. 'I'll kill you and then I'll kill the dog! Come on!'

Out on the pitch a uniformed police officer stood next to his German shepherd, which was itching to be let off its lead. I hurried over to the line of spectators. Judging by the crowd, this was the highlight of the day. The dog turned its head round and looked at its lead. The officer unhooked the lead from its collar and the dog bounded towards the hollering man. The 'criminal', who was the tall man in the boiler suit, threw his bag behind him and started running. The dog ran to the bag and picked it up with its mouth.

'Aaaw,' cried the crowd. The dog had let the criminal get away. Its handler arrived and hooked the dog back on to its lead and they walked off the pitch.

'Oi!' yelled the criminal. He wore protective equipment on his arm and was wielding a baton. 'I'll kill you and then I'll kill the dog! Come on!'

Another handler unhooked the leash and the dog ran at the criminal and sank his teeth into the material on his right arm. The dog maintained its grip as the criminal hit out with the baton.

'Good bite,' a voice said in my ear.

It was Clive.

'How can you tell?' I asked.

'It needs to grip the whole arm between its jaws and not just the sleeve. Problem is, doesn't look like the dog wants to let go!' He laughed. A number of people in the audience were also laughing. 'You see,' he said, 'new training regulations. They are ruining the profession.'

'Why is that?'

'I don't know if you heard about the incident with Essex police?'

I hadn't.

'Two handlers kicked a dog to death. Since then all the arts and flowers people have insisted on changes to the training regulations of the police dogs. It's silly because what those officers did was not part of the training regime anyway. What they did was wrong. Trainers are now only allowed to use positive training techniques and the dog is not allowed to be punished at all.'

'I see,' I said, wondering who the arts and flowers people were and whether I was one of them, and turning my attention to a third dog waiting to be let off its leash.

'Ha! That dog's taken a few whacks. Did you see it flinch?'

'Yes I did.'

'Yeah. That dog is from Slovakia, I think. A beauty. They don't have any of the ridiculous restrictions on training that we have here.'

'Is that why the police are importing dogs from the Continent?'

'Yes.'

'I've read that the police are having to pay translators to help them to communicate with dogs trained abroad.'

'Cor, what a bite.' He was watching the dog. It had the middle of the criminal's arm gripped firmly between its jaws.

'Down!' said the handler and the dog released its grip. The criminal was asked to walk towards the start line and the handler was told to keep his dog at his heel and follow behind the criminal. The dog looked as if it might lurch forward and attack the criminal again any moment and the handler had to continually re-enforce the dog's instruction to 'heel!'

'Our main problem is the new breed of dog handler. There is such a high demand for police dogs now that police officers who have never been pet owners are being given dogs. They have no interest in them and don't do the training exercises at home. As a result the dogs are simply less obedient and less effective in their tasks. It is the government's fault. They have no respect for professional standards. God knows what they are planning to do with all these dogs.'

The man work trials over, Clive bade me goodbye and left to pack up his stall. I made my way to the refreshments where the trophy winners would be announced. Elderly men were clustered round the scoreboard. Young uniformed men congratulated each other on their performances and were joined by their wives and newborn babies. Surveying the scene, I was approached by the company man.

'So I hear you are researching the legal regulation of dogs,' he said. I had mentioned this briefly in my application for membership.

'Yes.'

'I think that's great. It needs it. I have been saying that for a long time. I remember the first time I made a statement for the court and I thought to myself – an enthusiastic defence brief could make mincemeat out of it – if they knew anything about dogs.'

'Yes. What sort of cases have you given expert evidence in?'

'Arson.'

Clive came over and joined us.

'We were just discussing the need for regulation of dogs,' Derick explained.

'Oh don't talk to me about it. I spent a year on the committee convened to design the regulation of private security guards trying to persuade them of the need to regulate their use of dogs. They need training and a licence to stand in doorways but don't need any to carry a lethal weapon on a lead. It's insane.'

'And private security is taking over more and more police work,' I said.

'I know. I was called by a private security firm. Do you remember that our local police announced that they would no longer respond to burglar alarms because of the waste of police time?' Clive asked.

'Yes,' nodded Derick.

'Well this firm provided a service of responding to the alarms. They actually had an extremely successful clear-up rate – I think they managed to catch about eighty per cent of the burglars. Anyway, they called me because they wanted to set up their own dog unit. They were already using dogs to send into buildings that they thought had burglars in but they wanted to formalise what they were doing. I asked them what their dogs would do if they found a burglar. "Tear him to pieces," they told me. Well, I couldn't have anything to do with that and I told them so.'

An odd assortment of men, friends of Derick and Clive, joined us. After a series of short anecdotes about their 'bitches' and 'breeding' – they seemed to relish using the words – the club's chairman called everyone's attention and the award ceremony took place. When it was finished, the horse-faced woman came over and gave me her card. Her mannerisms were sprightly and she rocked backwards and forwards, hands in her front trouser pockets, as she spoke.

'Man,' she said, 'has been using dogs to hunt since time immemorial.' They were the obvious choice for the police, in her opinion. She was a member of the Kennel Club. Her interest in dogs wasn't a hobby, she said, correcting my attempt at small talk, but 'more of a sport'.

'The RAF dog trials are in September, if you fancy a good day out,' she said.

'That sounds lovely, Geraldine, although I am restricting my research to police work.' I wasn't about to make a habit of attending dog shows.

'Well, you know the police and the army are working more closely together these days. It might be worth a look.'

'Oh really? Well it would be nice to stay in touch in any event.'

'Please do.' She nodded her head vigorously. 'I might be able to take you to the Kennel Club for lunch.' She winked and strolled off towards her car, raising her arm in salute to various groupings along her way.

I phoned the taxi driver who'd brought me up and sat on the roadside to wait for him, waving goodbye to the men I'd met, as they drove past with cars loaded up with dogs and canine paraphernalia.

'Get in,' the driver called out the window as he pulled up outside the field.

'How was it?' he asked.

'Good, thanks. I missed the drug trials but it was an interesting experience for me.'

'What is your interest in the dogs exactly?' he asked.

I explained about the legal slant to my research.

'Is it just dogs you're interested in?'

'What do you mean?' I asked.

'Well, I was reading something 'bout sniffer-mice the other day.'

'Sniffer-mice?' I looked at him incredulously.

'Yeah. Apparently some American scientist has genetic-ally modified mice with super-sensitive noses. She says they're quicker to breed, cheaper to train and could work in smaller places than dogs. I s'pose they might be useful in some circumstances, but it sounds bloody stupid to me.'

I wondered if a Home Office lawyer had expressed concern that man might be a wild mammal for the purposes of the hunting ban, and had suggested finding a substitute for the dog.

'What do the police think of the idea?' I asked.

'I don't know. But the newspaper said that she's been invited to a conference here by the Ministry of Defence.'

The Ministry of Defence? A *conference* on sniffer-mice? This sounded bizarre. I had to find out more about this.

CHAPTER FOUR

The Fruitcake Burglary

I found the American scientist's contact details on the web and emailed her, asking what November's conference was about and when it was. She emailed me with the details of the organiser. The conference was on security applications of olfaction and would be held at the olfaction department of the MoD.

I sent the MoD an email explaining that I was a lawyer researching the use of olfaction in law enforcement and that I would very much like to attend their conference. Their response surprised me and I was unsure how to reply.

I phoned Tom and asked him to meet me for a drink. I told him it was urgent. He said he could meet me that evening, but he would have to be brief.

He was sitting in a booth when I arrived at the pub. He was tanned from the summer months and looked more gorgeous than ever. My heart pounded as I took a seat opposite him.

'What's up?' he asked casually. His eyes seemed to know how I felt about him and that made me nervous.

'You won't believe this,' I said, confident that I had good grounds for having called him. 'The Ministry of Defence has an olfaction department and they've invited me to give a presentation on the legal implications of olfactory surveillance at a security conference.'

'Well done!' he said. 'What's olfactory?'

'To do with smell.'

'Amazing. Sounds right up your street.' He didn't seem shocked to hear that there were military applications for olfaction.

'Oh yeah . . . like I am going to do it!' I laughed. 'I would be walking straight in to the Heart of my Paranoia.'

'Amber,' he reached across the table and removed a stray blonde hair from my forearm, 'you think you are a hippie. OK, your dad was a cannabis smuggler. But you are in fact a lawyer. And the world is not about Good and Evil. This is an academic conference and an excellent opportunity for you to publish a paper. You have to do it.'

I was shocked. I felt sure that the world of security was about Good and Evil. Could I really give a talk at the MoD conference? What was the inside of the Ministry of Defence like? I imagined a large grey bubble full of coloured buttons and scanning equipment.

'What are you worried about?' he asked gently.

'I don't know. I've never given a talk before.'

'What do you mean?' Tom said, getting to his feet. 'What about the talks you've being giving on dogs?'

'Well, I was nervous then.'

'But you did fine?'

'Yeah.'

'Look,' he said, 'I've got to scoot. I'll send you some papers that might help.'

The next morning I phoned up the conference organiser at the olfaction department at the Ministry of Defence and informed him that I would be happy to give a presentation at the conference. 'Who will my audience be?' I asked in what I hoped was a casual tone.

'Mainly scientists working in the field of olfaction and representatives of the British and United States military.'

'Right,' I cleared my throat, 'well I'll email you an abstract next week.'

'Great. We are all looking forward to meeting you.'

Shit. I looked up at my office colleague. He was absorbed in a telephone conversation, oblivious of my distress.

I spent the next week working solidly to ground my observations into a neutral legal framework of surveillance and procedural regulations. Much of the case-law on detection dogs was disheartening. Many courts had refused to accept that the use of a dog amounted to a search or that its deployment needed to be regulated. These decisions described the dog as 'an extension of the police officer'. Perhaps this was a valid interpretation. The question I asked myself was since when had a police officer been allowed to extend himself? Could portable cameras and listening devices be seen in the same light? Would a robocop be deemed a legitimate extension of the bobby? And if so, what basis would the court have for objecting when the Terminator turned up?

The reliability of the dog appeared to have been taken for granted as a matter of common-sense by the courts. In Devon in 1999, the police were called because a slice of fruit cake had been stolen. The police attended the house with a dog. The dog sniffed around the kitchen area and then appeared to follow a track some hundred yards away from the house. The dog stopped and indicated at an abandoned car in which a homeless man was sleeping. No fruitcake crumbs were found on the man or in the car. The man was interviewed and denied involvement. He said he had been sleeping rough in a barn. It was cold and he had found the car unlocked. He was convicted on the basis of the dog indication and inferences from his decision not to give evidence in court. In 2000 the Court of Appeal upheld his conviction for burglary.

Most of the academic material Tom emailed me on privacy and surveillance was written by 'lefties' and 'liberals' who made frequent reference to Orwell's *Nineteen Eighty-Four*. I played with this idea; there were olfactory references in *Nineteen Eighty-Four*, mainly to 'nosing zealots', that didn't appear to have been explored in this context by academics. I decided that the MoD might not be impressed with attempts to scare them with references to Big Brother, as he was probably a friend of theirs. Instead, I tried concentrating on the human rights implications from a governmental perspective.

The MoD informed me that equipment would be available if I wanted to make a PowerPoint presentation. I wasn't sure that I did.

'Visual imagery, that's the only strength of a PowerPoint presentation,' Tom had advised me on the telephone in the run up to the conference. 'They had a programme on it on Radio 4. A presentation which uses PowerPoint to structure the talk into bullet points is less likely to be understood by the audience than a presentation without PowerPoint. A presentation which uses it to spice up the talk with imagery, on the other hand, is more likely to attract the attention of the audience.'

I still wasn't sure I wanted to attract the attention of the security services.

'But what images could I use?'

'A picture of the Child Catcher from *Chitty Chitty Bang Bang*?'

I was touched that he remembered our first conversation. We'd talked about the imprisonment of children, and had got on to the use of sniffer dogs in schools. I'd compared the dog's nose to that of the Child Catcher who sniffs out children in Vulgaria for the ruling elite, which abhors their presence and seeks to imprison all of them. Pleased as I was with his suggestion, a vision of myself on stage inside a military compound with nightmare images from my childhood flashing on screens at the audience didn't appeal.

'Stored odours' nail China felons

BBC News
16 March 2006

A collection of deep-frozen human body odours is helping police dogs in eastern China track down criminals, the Xinhua news agency reports.

Some 500 odours, sampled from criminals and from different crime scenes, are stored in the scent 'bank' in Nanjing.

The archived scents are then presented to police dogs for comparison with odours recovered from fresh crime scenes, the bank's founder told Xinhua.

He said the process has so far led to the identification of 23 suspects.

Song Zhenhua said a specific odour would be added to the bank only after it had elicited the same reaction from at least three trained police dogs – a way of reducing the risk of accidental identification.

Mr Song said the samples are stored at a temperature of -18C.

'This way the scent sample is maintaining its "freshness" for at least three years,' he said.

The means by which Nanjing police capture the scents are not known.

CHAPTER FIVE

Hijacking the Yeast

The caterpillar pauses, feeling around in the air with its blunt head. Its huge opaque eyes look like the front end of a riot gear helmet. Maybe it's smelling him, picking up on his chemical aura.

   Margaret Atwood, *Oryx and Crake*

The morning of the conference arrived, and I was surprised to sleep through my alarm. I woke up with a jolt in bright sunlight. I hurriedly showered and pulled on black stockings, designer long skirt, black top, black Gucci suit jacket, black leather stilettos and scrambled around for a black handbag. Instead I found a fluffy orange and green bag and a variety of 1970s brown leather bags. I grabbed one, reassuring myself that 1970s items no longer represented Peace or Love but simply Fashion. I glanced in the mirror and pulled my hair back extra tightly, pleased with how stern I looked. I sprayed on some Dolce & Gabbana to complete the effect.

   I panicked a little on arrival at the station. The Way

Out sign took me nowhere. It pointed me in the direction of a wall. Feeling unsure of myself I walked up to it twice. I asked for further directions and eventually made it down the stairs and out of the station where a cab was waiting.

We drove through the countryside making amiable small talk for fifteen minutes until our arrival at the military compound.

'I'm not sure I can go any further love. I think it is private property. Not that it should be. It should be public property if it is owned by the state shouldn't it?'

'Do you think you should have a right to ramble through it and picnic there?' I asked.

The driver laughed. 'I wouldn't want to eat or sleep in there. I've heard they do all sorts of weird stuff in there involving animals and poisons. Best of luck though.'

I walked up to the guarded barrier where one of the guards pointed me in the direction of the 'Visitors' Room'. The room had about forty civilians sitting waiting in it. I wondered what their business there was but didn't find out. There was no queue at the reception desk so I went up and explained I was late for a conference on olfactory-based systems for security applications.

The woman behind the counter was reassuringly pleasant and smiled at my lateness. She telephoned through to the conference room and got no answer. She turned round, reached into a red tray, rummaged around for my name among otherwise identical clip-on passes and passed it to me over the counter.

'Oh good. There are others who haven't turned up yet.'

'Oh yes,' she laughed. She telephoned through to security and a guard promptly arrived to collect me. I briefly inspected my pass before pinning it to my jacket. It was a laminated piece of card bearing my name, the title of the conference and a photo of a dog's nose. This was the best piece of sniffer dog paraphernalia yet. We walked out of the visitors' room, past the barrier and along the road past rows of unmarked one-storey buildings. The guard wasn't very talkative. He showed me into one of the buildings. Coffee, biscuits and sandwiches were laid out on tables. He noted them filling my vision and pointed at the closed double door at the end of the room.

'It's started,' he said.

I opened the door slowly and stepped in, closing the door behind me. There were about thirty people in the room. Fortunately, they were all sitting facing the other end, where someone was talking. Immediately to my left by the door was a young well-built man, in a suit and tie, with a handsome face and jet-black hair. He smiled eagerly at me. I smiled back confidently but quickly and took a seat at the back of the row to my right. He immediately approached me.

'Amber?' he asked, and I nodded, a little taken aback, forgetting my name was pinned on my jacket.

He smiled and handed me a leather document holder and returned to his seat. I opened it up and was pleased to find a writing pad, pen and a glossy version of the conference brochure.

A Professor Harrison began his presentation with an anecdote about his own dog. 'She puts her nose into holes I wouldn't put a stick into.' He laughed, stretching open his moustache.

He didn't look like the sort of man who'd refuse to put his stick into many holes.

He gestured at the PowerPoint display showing a graph with a number of waves on it.

'These represent the "sniff rate" of a dog compared to that of a human. It is clear from this graph that a dog's sniff is more controlled than that of a human. We hope that further study of these sniff rates will allow us to develop an auditory cue to enable a dog handler to know when his dog is on to a scent.'

I was surprised that a conference speaker was willing to concede that a sophisticated device might be necessary for handlers to understand their dogs, what with the psychic connections between them.

Steve Tapper, a young stocky man with a crew-cut, was next to take the stage. He was studying canine detection in the United States. He opened with a reference to Aristotle, who had apparently said that smell was at the heart of perception. Tapper was concerned by the lack of peer-reviewed publications on the scientific validity of canine detection. Legal challenges to the use of dogs in the US suggested that the police might not be able to rely solely on the reputation of the dog's nose in the future. Judges and juries were starting to ask for scientific proof of a dog's ability to detect and identify specific scents. A lot of work needed to be done; the studies he had carried out had revealed a number of flaws in canine detection.

The United States was attempting to build up a database of human scents. I wondered if perhaps the Americans had the Stasi samples that went missing at the end of their regime. Scientists involved in Tapper's research felt confident that each individual had a particular scent that remained constant over time, regardless of diet or environmental factors. Scents had been collected from the armpits of volunteers. It sounded similar to the means employed to get the DNA database off the ground in the UK: armies of ill-informed volunteers.

The first impediment to the accuracy of scent as a form of identification was that the gauzes used to sample the scent were already impregnated with their own scents, ones shared by some humans. It had not yet been possible to produce an analytically clean medium to impregnate with a scent.

I could see why this might be an impediment. Imagine if knives carried their own DNA strands that were identical to those of some humans or if window-panes were made up of markings identical to fingerprints. It would make me feel a little insecure in a criminal justice situation. I followed this train of thought. Maybe that was the idea . . . maybe they'd cut a deal on a guilty plea . . . I guess the police could just choose to ignore the dog's bark when it's doing it at an inanimate object. If I am ever busted, I thought, I'll claim I am a vase and get myself a dog as a witness.

'Ironically,' he concluded, the gauzes used by the FBI, manufactured by the family friendly Johnson & Johnson company, were the most heavily contaminated gauzes.

The second impediment was the absence of an agreed unit of measurement. Olfaction lacked a scientific language. He referred us to the website of the Scientific Working Group on Dog and Orthogonal Detection Guidelines. They were working on this problem.

Someone on the front row sneezed. 'Bless you,' a red-haired and voluptuous American woman called out across the room.

After a short talk from the doctor who had discovered, accidentally, how to increase the olfactory sensitivity of mice, the time arrived for refreshments. I was shy and wary of my companions, many of whom were crowding together awkwardly round the coffee. I decided to go outside and have a cigarette. I would find a small number of smokers (if there were any) easier conversation.

The woman with the glossy red hair was the only one actively mingling. She was very tall with an imposing figure and seemed to already know a fair number of people there, including the man who had greeted me, Dr Teckler. She had a large greedy face that gave her some sex appeal. Her badge said she was a doctor from the US government. Her name was Debra.

There was only one other young woman in the room. She was short with curly blonde locks. Her name was Lucy and she was from a private company. We exchanged smiles.

I made my way outside and lit up a cigarette. A man who had been sitting behind me during the talk ambled out towards me. He didn't look like a smoker.

'Randy Oldensaw.' He outstretched his arm. 'And you are?'

'Amber.' I held his hand and smiled. His badge read 'US Government' and I asked him what department. 'Defence, in Washington. Your badge says CBA, what is that?'

'Criminal Bar Association.'

'Did they send you here?' His tone was aggressive. He didn't appear to like whoever had sent me.

'No. I am doing my own research in this area.'

He looked confused and I wanted to change the subject.

'So, do you think the mice will catch on?' I asked. I'd been surprised by the lack of enthusiasm shown for the mice when the scientist whose work had led me to this conference gave her talk.

He stood still with his hands in the pockets of his cream trousers and his feet hip-width apart and wobbled his face from left to right a couple of times, like a bloodhound. 'No,' he said, lifting his head and looking up at the unusually blue sky. 'It's an image problem.'

With a tilt of the head, his eyes, buried between a large forehead and dark grey bags, peeked out at me from either side of his prominent nose. 'I couldn't ask one of my men to walk round with a mouse,' he said in a sincere and concerned voice. 'The same impediment applies to pigs, and they have great noses.'

'Yes, the German police had to abandon their use of pigs for that reason, didn't they?'

He shrugged.

'What about cats?' I asked.

'Well sure, but have you any idea how difficult it is to train a cat?' He shifted from one foot to the other, almost dancing around me. I was beginning to understand police officers' love of dogs. Dogs are unconditionally obedient and cats tend to be more independently minded.

'But the Russians did it,' I said, remembering something I had read. His body muscles visibly contracted and he froze and stumbled forward.

'What do they use them for?'

'Detection of caviar smuggling.'

'Ohhoh, well yeah.' His body relaxed as he spread his legs apart again. 'Problem they'll have is the cats will eat the caviar and they won't have any evidence left.' He must have noticed I wasn't impressed with his answer and added, 'I trained my pet cats, just for a laugh at home, with laser pens. They got pretty good.'

I'd finished my cigarette and told him it was probably time for the next talk. He followed me in and we returned to our seats. In the corner of my eye I could see him reading my abstract in the conference brochure and a couple of minutes later he tapped me on the shoulder.

'Do you really think privacy is going to be a problem for olfactory technology?'

'I don't know, but it could be.' I was desperate to appear neutral on the subject. I was secretly afraid of the US government. I didn't want it to decide I was an enemy.

'Really? How could it be?' he probed.

'Well, it's context dependent isn't it? It's related to one's expectations of privacy. You wouldn't be unpleasantly surprised if required to pass through a metal detector before boarding a plane. If you were approached by security personnel carrying a metal detector up to you in the park where you are enjoying a picnic with your family, you might think of it as an intrusion on your privacy.'

There was a pause. He appeared to be thinking very carefully about what I had said.

'Actually,' he said, his brows unfurrowing, 'I wouldn't mind. So long as they were doing it to everyone. I guess that's it. Lack of discrimination. That's what's important.'

This struck me as an unorthodox solution to discriminatory policing and I found it difficult to formulate a response before the next speaker summoned our attention.

Scientists in France claimed that they had succeeded in 'hijacking the yeast pheromone pathway' to install genes from mammalian olfactory receptors within it. Images of the yeast changing colour in response to different odours were shown on the PowerPoint display; cocaine running around its new brain and it started flashing green. Things would have come to a pretty pass if I couldn't trust my mate Marmite not to grass on my drug habits.

At the next coffee break I found myself talking with Lucy and an eccentric-looking Hungarian professor who had designed a school of sniffer robots. He had never talked at a security conference before. Both Lucy and I found him amusing company. He and I were next on. I told him I was a bit nervous and he said he was too. I worried about my own presentation throughout his talk, my eyes resting on robots trailing scents across the screen. I hadn't prepared a PowerPoint presentation, thinking that, as a lawyer, I could get away with a frill-free speech from behind a podium. No one asked any questions and it was my turn to speak.

'The most important concept we need to understand in order to grasp the potential human rights implications of an expansion in olfactory surveillance is the right to privacy.'

I locked eyes with two young insect specialists in the front row.

'The right to privacy has been said to be at the heart of liberty in the modern state. The right to privacy implies that in the absence of compelling justification, we should all be free to move about without fear of being systematically observed by agents of the state. In accordance with the right to privacy, the law requires the police to have reasonable grounds to believe that a crime has been committed by a person before searching them. Until relatively recently, the police had to physically search a person's pockets or houses to find out what was inside of them. This is no longer the case.

'Some courts have dismissed the argument that a "sniff" constitutes a search as absurd, on the basis that mere sensory perception cannot infringe the right to privacy because odours emitted from a person are exposed to the plain perception of the public at large. This reasoning ignores the fact that dogs are used precisely because of their supposed ability to detect odours, which are not exposed to the plain perception of the public at large.'

I looked briefly around, attempting to gauge their response. They appeared to be listening attentively.

'The US Supreme Court has recently recognised the threat to privacy posed by new surveillance techniques in *Kyllo v. United States*. In that case the police had aimed a thermal-imaging device at the appellant's residence to detect heat emanations associated with high-powered marijuana-growing lamps. Based on the thermal-imaging information, police obtained a search warrant for the residence and found marijuana. In the appeal against the resulting conviction, the court held that when the police obtain by sense-enhancing technology any information regarding the interior of the home *that could not otherwise have been obtained without physical intrusion* into a constitutionally protected area, that constitutes a search and reasonable grounds are required to make that search legal. The court observed that this would assure preservation of that degree of privacy against government that existed when the right to privacy was adopted.

'There are some judges who believe that if constitutional scrutiny is in order for the imager, it is in order for the dog. Some courts have declared that the use of a sniffer dog, without reasonable suspicion to justify it, amounts to an illegal search and that any evidence found as a result of it should be declared inadmissible.'

Having dealt with the law on privacy, I briefly discussed the difficulties presented by the use of olfactory evidence. I threw in an account of a challenge to the use of dog evidence that had recently been made in the Court of Appeal on the basis that the dog couldn't be cross-examined, earning me a guffaw from the senior member of the olfaction department. I argued, in line with recent authorities on ear-print identification, that until scientists were in possession of a database of different odours, and could show that the dog was reliably picking up on the unique odour of the person or object in question, it shouldn't be admissible.

'Any questions?' the chair asked.

A man in the front row put his hand up. He was in his mid-fifties and wore silver-rimmed bifocals, the kind that droop stylishly under the eyes. He held his chest upright, broadening it while allowing his lower body to sink into his pelvis, which rested at the front of his chair. 'Presumably what you are saying about intrusion on privacy and legality applies to a lot of the new technologies we are developing.'

His accent was soft and American.

'Probably.'

'And are you saying that if we suspect someone is a suicide bomber, we invade his privacy without any prior suspicions, and we find bombs, that we would not then be allowed to rely on the evidence because of the privacy breach?'

'Well, no. In practice, and in those circumstances, that evidence would be permitted despite police having broken the law to obtain it.'

Randy shouted out from the back row. 'It's not that simple though Amber. If we think someone is carrying a bomb, we don't stop and search the guy, we terminate him.'

'Well, that's very exciting, but what is your point?' I snapped, and instantly regretted doing so. He was actually making a very good point. If the science of olfaction was not 100 per cent accurate, it probably wasn't the best tool for deciding which people to terminate. I hadn't realised that was on the agenda.

'Well, no, I'm just saying,' he replied. 'That is what we would do.'

Fortunately, there were no further questions and so I collected my papers, stepped away from the podium and started to make my way back to my seat, when the man who'd asked the first question put his arm around me and peered into my face. The understated mullet softened his thick neck with short grey ringlets and the freshly laundered fabric of his silk sleeves exuded a sweet and comforting scent.

'Oh boy,' he said smiling. 'Now we really hate you.'

'Why?' I smiled as breezily as possible.

'We can't work out what side you are on.' I maintained my smile, not knowing what side to name, and proceeded to the back row where Randy was waiting for me, an affectionate and expectant expression on his face.

'I'm sorry about that,' he said. 'I was trying to find a polite word for kill.'

'That's OK,' I said.

'You see,' he stood before me in his blue cotton shirt and cream coloured pants, 'we have this new technology called Terrorhurts.'

I laughed. 'Terrorhurts?'

He looked at me quizzically. The black pupils of his eyes constricted momentarily before refocusing on the letters of the new technology. 'T-E-R-A-H-E-R-T-Z.'

'Oh. Did you pick the name because it sounds like terror hurts?'

He repeated his quizzical look and then shook his head. 'Oh, I hadn't noticed that. No. It uses soundwaves, which are measured in hertz. If you point it at someone they freeze still rather than dropping to the floor. Actually, it's quite invasive. Really messes with your internal organs. It's not used as a weapon yet though. At the moment it is being used to detect weapons by looking through clothing. It doesn't harm the person when it is used in this way but it could count as an invasion of privacy.'

'Yes, it sounds like it.' I moved away from him to get myself some water. Dr Teckler sidled up to me along the wall on my way out.

'I am sorry about that, Amber,' he said glancing over at Randy. 'We don't really like the American attitude to some things, like instant termination – it can be very difficult for us.'

I didn't really understand what he thought there was to apologise to me for, but I nodded appreciatively and got myself a glass of water. I was excited about the next talk as it was to be given by Professor Hall, a senior scientist at the Police Scientific Development Branch, which is part of the Home Office. The role of his department was to develop technologies for assisting the police in their fight against crime. The government had recently established a dedicated funding stream to focus science and technology attention on crime reduction and I was looking forward to hearing what the money was being spent on.

To my surprise, his talk was on bees. Sniffer bees.

CHAPTER SIX

The Sting in the Tail

Before they had been beasts, their instincts fitly adapted to their surroundings, and happy as living things may be. Now they stumbled in the shackles of humanity, lived in a fear that never died, fretted by a law they could not understand . . .

H. G. Wells, *The Island of Dr Moreau*

Bees, Professor Hall enthused, had been around for two hundred million years; 'they survived what dinosaurs could not.' Dogs had only been around for five million years. There was massive redundancy among the insect population and the Home Office was looking into employing the bees to sniff out illicit substances. This man now had my full attention.

He explained that bees were captured and deprived of food for a short period of time. He seemed anxious to stress the brevity of this period, seeming fearful that some members of the audience might be insect rights campaigners. Maybe he thought I'd get on an insect cruelty trip.

The bees were placed in plastic holders. Lucy, or Bee

Girl as I liked to think of her after learning that her company specialised in sniffer bees, leaned over and whispered in my ear. 'Actually we use sellotape 'cause we haven't found any appropriate plastic holders yet.'

The bees were subjected to a constant air flow. Air was re-directed over the target material and the bees were simultaneously fed sucrose solution.

'By this process they are trained to associate the smell of the target material with food.' Their response to the association is to wag their tongues ('proboscis extension' is the technical term).

A portable unit had been created to enable the bees to be used at airports. One such device was being used in a Paris airport. Cameras were installed inside the device with three bees. Security officials pass an item, a suitcase for example, past the bees. They then watch the bees' tongues on TV monitors. The beauty of the device, he explained, was that it looked like a box. Only the security officials knew it contained bees.

'What is our final goal? Free-flying bees. I would like to see trained bees being sent out into the open. We will be able to rely on their natural behaviour, which is to inform each other when they smell food. They will swarm in on drug laboratories across the country.'

Christ, I thought. This man seemed mad. He wanted to transform the Fuzz into the Buzz. The future of sting operations looked menacing.

One disadvantage to the bees was they were not very good in the winter, he conceded sadly.

His formal presentation over, he concluded with a personal observation that some people might be bothered by the presence of sniffer bees around them.

'It can be irritating having an insect flying around you, as we all know. But that is nature.' He shrugged dismissively.

Lines from *Brave New World* flew through my mind. 'The air was drowsy with the murmur of bees and helicopters . . . what man has joined, nature is powerless to put asunder.'

'Any questions?' asked the chair.

'You say they are ninety-five per cent effective. What about false positives?' asked the man in the grey linen trousers who couldn't work out what side I was on.

'Well they do sometimes wag their tongues at substances other than those on which they are trained. This hasn't been factored in to the effectiveness rate. But they don't do it a lot. They are pretty good.'

'What about nectar?' he pressed.

'We haven't tested them on nectar. But we have tested them on washing-up liquid and castor oil and they rarely react to it. They are not a silver bullet,' he conceded, 'but they give you something when before you had nothing.'

It was time for lunch. I drifted out to the room with the sandwiches and helped myself to a few. A doctor from Porton Down, the secretive science park engaged in biological and chemical weapons research, approached me.

'Hello. My name is Paula. I thought your talk was extremely interesting. I work on biological weapons detection. We have no idea if our evidence is going to be good enough to stand up in court. Can I have your contact details? I want to ask my boss if she'll commission you to do a study on the forensic validity of our detection methods.'

Surely they had lawyers for this sort of thing, was my first thought. It felt like an invitation to the Dark Side, which I didn't expect to be followed up by her boss, but I was flattered nonetheless and gave her my details. Perhaps I had pulled off making the presentation without striking them as a woolly-headed libertarian. In fact, these people appeared to like me. With my increased confidence I relaxed and stopped worrying about my persona.

A sixty-year-old red-faced man, whose belt had given up on holding either him or his shirt in place, came and joined us.

'You said it was difficult to cross-examine a dog. What do you reckon on the likelihood of cross-examining a bee?' He laughed through a mouthful of sandwich.

'Well you laugh at the possibility,' I replied, 'but you know it wasn't all that long ago that we prosecuted animals in criminal courts and relied upon them as witnesses.'

'Really?'

'Yes. Bartolomew Chassenée, a distinguished French jurist of the sixteenth century, made his reputation at the bar as counsel for some rats. The rats were charged with feloniously eating up and wantonly destroying a barley-crop and were ordered to appear at the tribunal. Chassenée was their defence lawyer. In view of the bad reputation and notorious guilt of his clients, Chassenée had to try every legal loophole he could think of to ensure their acquittal. He eventually succeeded in excusing the non-appearance of his clients on the grounds of the length and difficulty of the journey and the serious perils that attended it, including the unwearied vigilance of cats, their mortal enemies that lay in wait for them at every corner and passage.'

'You're kidding?'

'No. And at least one man was acquitted as a result of testimony of his cat, dog and rooster. So who knows what the future holds.'

He laughed, not knowing how seriously to take me.

The talks were over and it was time for us to get into separate groups for a brainstorming workshop session. Each group was made of ten people and had a separate task. My group included Roger, the man who'd pulled me aside and said he hated me, and Debra, Bee Girl and the Hungarian. It was chaired by the youngest of the MoD organisers, who though undoubtedly older, looked liked a boyishly handsome eighteen-year-old head of a school debating society. Our task was to design an artificial olfactory device for the detection of illicit substances in half an hour and report our findings back to the group. This was, ethically, a difficult position for me to be in.

Fortunately, like any workshop, it was a rambling mess of pointless idiocy and I had nothing useful to add.

Eventually, the chair took control.

'OK. First we have to decide what setting we are designing this apparatus for.'

'Somewhere with large crowds, like an airport,' said Debra, waving her hands excitedly in the air.

'Yup,' nodded the chair, typing on to his laptop. 'So, an olfactory detector in the design of a metal detector that the individual would be filtered through.'

'I don't know,' said Roger. 'I think maybe we should shape it by its distinctive technological capabilities rather than copying designs from other technologies. Maybe concentrate on the target substance at the point of danger. Perhaps small vents inside the aeroplane itself. That way the device can operate effectively in an isolated area and the plane wouldn't be able to take off until the device gave the all clear.'

I couldn't help but voice my approval of what struck me as a sensible idea, suitably proportionate to the risk. The others didn't like it though. They were thinking big. They were thinking everyone . . . no rationale . . . no discrimination. They were thinking Crowd Control.

'One problem we have to consider is that of Time Delay,' said the chair. 'The artificial olfactory system may take some time to clean itself out. Unlike a dog, it can't immediately cleanse its system by blowing its nose.'

'Oh God!' cried Debra. 'I have just had an evil thought.'

We all turned to look at her. 'Forgive me people. What if . . . Oh God.'

She grabbed both sides of her head, pressing the palms of her hands into her temples. Her red fingernails stuck up above her head. She looked like she was metamorphosing.

'It is an evil thought but what if . . . Oh God . . . what if the terrorists worked in a pair . . . Oh God! I am sorry people for having these evil thoughts. But what if they deliberately work in a pair so that the first to go through the olfactory detector would set it off with his explosives, clogging up the system, allowing the second terrorist to go through with his explosives undetected?'

Surely this was what was meant by the 'the problem of time delay'? She wasn't having a miraculous epiphany.

'She's right,' someone said. 'We need more dogs until we can fill the gaps.'

What gaps? I thought. Our workshop had invented a technology, which didn't work, and suddenly we had to find substitutes for it.

Everyone agreed more dogs were needed.

'We could use dogs and bees in the airport,' said Debra in an authoritative tone and turned to smile inclusively at Bee Girl. I watched Bee Girl's blonde curls enter my right eye vision and shake.

'I don't think you'd really want bees flying round an airport,' she smiled leaning forward.

Debra flicked her hair and crossed her legs. 'Well I just thought, you know, we could have the dogs stationed in one area, and the bees covering the general area, because they can fly.'

I whispered a reminder to Bee Girl that Professor Hall planned to help the Home Office send bees out into the city on drug detection duties.

'I don't really think that's realistic,' she whispered back.

'Maybe,' Debra resumed, 'we should consider genetically modifying dogs to meet our requirements.'

'Flying dogs,' the Hungarian chuckled to himself.

'Yes!' Debra exclaimed, overhearing him.

Roger was thinking deeply about the difficulties ahead. He sat leaning forward and out of his chair, arms poised ready to leap on crossed legs. 'The problem we face in the United States is that we have run out of working dogs. The only dogs left in the US are family pets.'

Everyone except for Bee Girl looked concerned by this. I had obviously failed to appreciate the urgency of the situation; there wasn't time to breed more dogs? How many did they think they needed? I'd read that in the First World War, Germany had conscripted thousands of Alsatians into the army, using force to take those that were treasured family pets. I decided not to mention this. These people might be crazy enough to copy the idea. In fact maybe that was what had happened to the nation's missing dogs. Could the Russians have taken the cats?

The official, in his exasperation, had written 'Other than dogs' on the PowerPoint. Our half hour was nearly up and he asked us to leave him to write up something to present to the other teams. He looked stressed.

'Perhaps you should help him,' I said to Debra as we all got up to leave.

'I'm going to,' she said, making her way over to him.

After the end of the workshop presentations, Jeremy, the senior member of the MoD olfaction team, presented a summary of the conclusions of the three workshops. Bees and moths had complex brains and could fly. Dogs were easier to train than moths. Microbes didn't have brains and were ideal for genetic manipulation. Perhaps flying dogs were a good idea, he joked, though even better might be to 'clone dog-faced policemen'. Artificial e-noses were good for learning how brains worked but they lacked the selective amplification capabilities of the mammalian system in that they got side-tracked by other substances. The United States was already having difficulty finding sufficient working dogs and was having to import them. There remained the possibility of using birds and fish but for now the dog remained the benchmark for odour detection technologies.

The conference over, Dr Teckler walked me to the barrier and I asked him how he came to be involved in olfaction. He told me he had joined the Ministry of Defence shortly after graduating. He said he had been lucky, as there weren't many jobs for scientists these days. His work was mainly with police and military dogs.

'I really enjoyed your talk,' he said, turning his kind blue eyes to face me at the gate. He was quite handsome with a square jaw and strong chin. 'You are the only lawyer in the UK researching this area. It would be good to keep in touch.'

'Thank you very much for inviting me. I've really enjoyed myself.'

'Are you OK to get back? Why don't we see if you can share a cab with Roger?'

He walked me past the gate to where Roger was waiting for a taxi that promptly arrived. Roger opened the back door, took my hand and guided me in. We shared an animated conversation about the literature of smell.

'There is quite a bit on smell isn't there?' I said.

'Yeah, I guess there probably is. I wouldn't be surprised if some guy hadn't Googled them all and made a list of them on a website.'

'I wish they had. Have you read *Perfume* by Patrick Süskind?'

'No.' He parried. 'Have you read *Timbuktu* by Paul Auster?'

'No.'

'It's a novel written from the perspective of a dog. And there is another book where the guy is driving round with his dog and they pick up a hitchhiker and the dog senses the guy is evil?'

'I don't know it. The man in *Silence of Lambs* has a powerful sense of smell doesn't he?'

'Oh, yeah right. I think the children in *Peter Pan* smell danger too.'

'And Jack and the Beanstalk of course. "Fe fi fo fum, I smell the blood of an Englishman!"'

'And Marcel Proust of course.'

I frowned in ignorance.

'They call it the "Madeleine effect". Marcel Proust is best remembered for his *Remembrance of Things Past,* a huge and influential novel of seven volumes in which he discusses his preoccupations with the "lost landscapes" of memory which lie beyond voluntary recall. You know madeleines, the sponge cakes you get in Paris?'

'No.' I was embarrassed to be so outcultured.

'Well anyway,' he patted my leg affectionately, leaning back against the background of countryside in the rear window. 'This little citrus-smelling madeleine brings back all the rich, sensual background of the village of Combray, where the narrator spent his childhood. Proust called odour "a nightlight in the bedroom of memory".'

I thought we were getting on surprisingly well and ventured to ask him his job title. He explained that he was not part of the US government, as was stated in the conference brochure. He was working on olfaction for the US government on a freelance basis. He was a consultant. His present job was to find out who was working on olfaction and provide their details to the US government.

'So you've been going to a lot of these sorts of events?' I asked.

'Oh yeah.'

'And have the olfactory capabilities of many other animals been researched?'

'Yup. Dogs are facing sniff competition,' he joked and then proceeded to overwhelm me with the research he'd collated.

Computers were training rats to stand on two legs when they smelled a cocaine mimic. Machines above their heads only dispensed cocaine-flavoured food when a sensor on their back detected them standing up. Once trained, sensors were attached to the rats and alerted the controlling computer to a rat's location, so it could monitor the exact whereabouts of any contraband.

In another project funded by the Pentagon, University of Florida scientists had identified the neural signals rats generated when they had found a scent that they were looking for. Each rat had electrodes permanently implanted in three areas of the brain. Trained rats were sent to forage for their target smell, while the electrodes allowed researchers to identify the brainwave patterns associated with finding that smell. Signals from the rat's brain could be relayed to a radio transmitter pack strapped to the animal's back. Some rats had been trained to detect human scent.

'The pleasure zone of the rat's brain is stimulated by the scientists every time the rodent gets a whiff of human odour, which in time trains the rat to seek out such smells for a reward.'

Sniffer moths were being trained by US scientists to detect marijuana. Previously, insect experts had believed moths only identified odours by instinct.

'In nature, a moth must detect the tiniest whiff of sex pheromone from a mate who might be over two hundred yards away and scientists are learning to take advantage of the insect's biological traits.'

'What about the problem you identified in your question to Professor Hall about bees continuing to react to nectar?'

'Well, Pentagon scientists claim they have succeeded in getting their bees to ignore flowers and swarm around explosives. I don't think instinctive reactions are as difficult to manipulate as once thought.'

'Gosh. And how do moths make their indications?'

Roger puckered up his lips and made a funny sound. 'They suck. Here,' he said, reaching into his back pocket, 'I'll give you my business card. I can send you a database of future conferences in this area.'

'Thanks. That's very kind of you. What is the US government going to do with the contact details you provide them with?'

'Knock on their doors and give them money for their research.'

It seemed like his employers had some big plans for the science of olfaction.

It wasn't until we parted company that I realised that all my cultural references had been to stories in which the powerful nose belonged to an evil spirit and was used to harm the innocent; and all his stories seemed to feature heroes with powerful noses rooting out evil or poetically bathing in the beauty of odour. Could he have decided what side I was on from that?

Tails of the City

The Economist
29 October 2005

The mayor of New York's latest annual report on city management is a glowing paean to the rising quality of life under his rule. Crime is down 5% year on year, violence in jails is down 25%, civilian fire deaths have fallen to the lowest level since 1919 – all good news with an election due in November.

The big exception is complaints about rats and mice, which rose 40%, to almost 32,000, in the year to March. Officials blame the jump in reports on the ease with which complaints can now be made through a new City hall telephone hotline. But many New Yorkers would beg to differ. Anecdotal evidence suggests more rats are being seen in parks and on subway platforms – where, at a foot in length and a pound or more in weight, they are hard to ignore . . .

CHAPTER SEVEN

The Witch's Sense

Smell is the witch's sense, sniffing out the spirit of what has been, detecting an essence after the fact of its existence. It is the formula for time travel, lingering on for decades as the scent of cedar in an old sea-chest.

Lyall Watson, *Jacobson's Organ and the Remarkable Sense of Smell*

Ensconced in the British Library, I was finding the sense of smell intriguing. Despite centuries of research, scientists still didn't understand how it worked. Its secret remained elusive, and contained, according to several commentators, all the mysteries of the universe. Some scientists said it was only a matter of time before these mysteries would be unravelled by technology. Others said it would continue to elude us because odour was not a physical property, but existed only in the mind.

I darted manically across the green carpeted floor, between the catalogue, collection point and reading desks of the Humanities room, digging out every book and paper I could find pertaining first to smell, then to odour, odor, osmics, osmology, osphresiological, olfaction and olfactory. I rifled through them for clues. The pile of books and journals on my desk grew bigger as the faces of those around me faded out of sight. Every now and then I looked up at the Santa Claus look-alike a few rows down from me. His eyelids played restlessly behind half-moon spectacles and now and then he dozed off contentedly as if the lamp above his desk was his fireside.

I zoomed in on a handful of recent journal articles and what struck me as the most informative books: Lyall Watson's *Jacobson's Organ,* William McCartney's *Olfaction and Odours* and Dan McKenzie's *Aromatics and the Soul.*

The use of smell in human culture had a long and strange history. The Devil traipsed through it accompanied by a 'sulphurous stink' and perfumed saints pin-pointed sins of the flesh by their odour. Bad smells identified witches and warlocks. There was even an eighteenth-century Bill of Parliament to protect men from the 'witchcraft' of scents that could manipulate their minds.

In the beginning of the twentieth century, according to McCartney, the 'odour of sanctity' was translated into the formula $H_6 H_{12} O_2$ and, according to McKenzie, smell was simultaneously recognised as:

> the most subtle of all our senses, possibly, as some hold, because of the ancestral appeal to our (more or less repressed) animal nature. So subtle is it, indeed, that . . . its stimuli may not, on occasion, emerge into consciousness at all. They remain below the threshold. So that, although subjected to their influence, we may remain ignorant of the cause of that influence.

Our olfactory receptors are directly connected to the limbic system, the most ancient and primitive part of the brain, often referred to as the reptilian brain, which is thought to be the seat of human emotion.

Emotional scents were said to be catching and the ecstasy of churchgoers apparently blends with the sexually scented clouds of incense to unite congregations in peaceful harmony and contemplative bliss. In the science of smell, suggested McKenzie, it may be possible to find:

> one explanation of the mysteries of crowd psychology of those unreasonable waves of passion that sometimes sweep through masses of people and lead to all manner of strange happenings, like crusades and holy wars; autos-da-fe; witch burnings; lynch-murders; state-prohibition; spiritualistic manifestations; and other miracles.

The sensitivity of animals to fear in humans was widely acknowledged in the academic journals I'd found and recent research showed that women could discriminate between armpit swabs taken from people watching 'happy' and 'sad' films just by sniffing them. When the emotional odour secreted by rats in receipt of 'a mild but frightening electric shock' wafts into the cages of well-treated rats, the pampered rodents begin to panic. So there was a smell of fear. And humans, as well as animals, were able to communicate and detect it, though we may release our messengers involuntarily and receive them subliminally without conscious recognition.

I discovered that the ability to detect chemical signals was probably the earliest type of sensing among primitive organisms and that smell is the principal chemical sense. Lyall Watson illustrates how all living things continue to subscribe to a universal system of chemical communication. When an African killer bee stings it releases an alarm that smells like bananas, causing the other bees to become agitated and sting in their turn. Plants under attack from viruses release chemical signals warning their neighbours to erect defence barriers. The *Drakaea* orchid produces a chemical that is identical to that produced by a certain wasp to trick it into thinking it will find a mate inside its petals. The *Lophiomys* rodent uses scent to unnerve its opponents by releasing a powerful odour to make its enemies feel dry-mouthed and distinctly uneasy; 'it seems to do this with the help of a volatile, invisible mist . . . It knows where we live and it goes there, arousing primeval concerns. It evokes unreasonable unease by reaching out to parts of our brain that deal with such misgivings.' Sharks detect the scent of distress as easily as blood, and zigzag round it before homing in on their frightened prey.

The question of how an organism determines the relevance of an incoming odour remained a matter of debate; they either learned through experience or they were genetically programmed to recognise the signs. Experiments suggested that baby rats are born with an innate fear of adult male rats and that they lose this fear after around fourteen days of age, when they are no longer in danger of being eaten by them. Humans were thought to be born without odour aversions or preferences and it is believed that we learn to recognise the scent of our mother's nipples through experience. But research was ongoing. Psychologists had recently shown that humans could be conditioned to fear insignificant odours. In an experiment a neutral odour, unnoticeable by the subjects of the experiment, was released into a room while they underwent a stressful test. When the subjects were presented with the scent on a subsequent occasion, they exhibited signs of stress in an otherwise relaxing setting.

McKenzie noted the ability of persons with instinctive

dislikes for certain beasts to detect their presence before catching sight of them. He cites this newspaper letter:

> I myself loathe and fear spiders – so much so that I have been known on more than one occasion to go into a darkened room and to declare the presence of one of these creatures, my pet abomination being subsequently discovered

and asks whether this phenomenon might be explained by spiders throwing out a scent that bristles the hairs on the back of the neck of those who know and fear it.

On glancing up from my desk I locked eyes with a surprisingly handsome man immediately opposite me. I had yet to master the art of smiling at these moments and dropped my head.

According to the late scientist and essayist Lewis Thomas, smells tell us:

> When and where to cluster in crowds, when to disperse, how to behave to the opposite sex, how to ascertain what is the opposite sex, how to organize members of a society in the proper ranking orders of dominance, how to mark out exact boundaries of real estate and how to establish that one is, beyond argument, one's self.

The man with whom I had locked eyes was now engaged in retinal flirtations with a young girl in the row next to me.

I found accounts of people permanently losing their sense of a smell as a result of 'exposure to a very strong and disagreeable odour'. Olfactory hallucinations appeared to be widespread.

A loud speaker announced that fifteen minutes remained

for those of us who wished to order any books. I made my way to the workstation and punched in a list of items I'd come across in the course of my reading, collected them and added them to my desk.

In my haste I accidentally ordered a non-smell-related article from 1940s America that put a new twist on military applications of scientific research into the functioning of sensory organs:

> How much noise can the ear stand? . . . newspapers said that if Disney had turned on the whole of *Fantasia* at once he would have killed the audience, but the thing was never tried out. Fernberger's committee complained that we have absolutely no knowledge of what happens under such circumstances.

I wanted to know what the Fernberger committee was. And why they wanted to know whether it was possible to kill a cinema audience. But it was too late to order any more material. The document in my hands was the abbreviated version of a paper read before the Eastern Psychological Association by Edwin Boring, some four months after the United States declared war on Germany and Italy.

> You want to know, of course, *the speaker said,* how Perception is going to help the United Nations win the war, or rather how Psychologists are going to help Perception to help the United Nations to win the War. It is a reasonable wish, which I am not going to be able to gratify. In the first place, no one person knows about all the perceptual research in progress, except perhaps the voiceless solipsistic files of the National Defence Research Committee. In the second place, what is known

can not be told. Most of you have come up against this silence of classified communications and materials – the restricted, confidential and secret classes which research workers must swear to protect. War, we find, reverses the terminal scientific process of publication. In-sights and discoveries that have defensive or offensive value must end, not in a journal nor in a speech like mine, but in the secrecy of selective communication . . . Truth is no longer public property. It has to be made to work for the right side. For much of what is going on both you and I must be content to wait until after the war.

Surveying the pile I had accumulated on my desk, I glanced over at the clock behind the enquiry counter and realised I didn't have long to get through it before the library closed. As interesting as the beliefs of the Ongee people in the Andaman Islands were, particularly the idea that 'spirits, who wish to regain life, constantly seek out the odors of the living and cause their deaths by absorbing all of their smells', I had to put them to one side. With Roger's comment about the nightlight in the bedroom of memory sweeping through my mind, I scoured through the documents on memory and odour. McKenzie outlined the 'backstairs influence' of odours upon the mind, describing how they break in upon our privacy, bypassing our cerebral hierarchy to arouse memories independently of the will. J. H. Kenneth's 'Pamphlet on Osmics' recounted an experiment in which volunteers were asked to inhale odours and then to describe their experience. On being subjected to the smell of camphor, a man was overcome by the memory of a forgotten incident from his childhood, thirty years beforehand, when he had been shut up in his mother's wardrobe as a punishment: 'camphor first produced a feeling of mental distress, next the visualisation of a wardrobe door, then the image of the wardrobe as a whole, attended by a sense of suffocation, fear of the dark and recollection of kicking and hammering to get out.'

A surprisingly large amount of material on odour and memory dealt with Jewish experiences of the Holocaust, including Olga Lengyel's account of the appearance of Irma Griese, an exceptionally beautiful SS concentration camp official:

> Wherever she went she brought the scent of rare perfume. Her hair was sprayed with a complete range of tantalizing odors: sometimes she blended her own concoctions. Her immodest use of perfume was perhaps the supreme refinement of her cruelty. The internees who had fallen to a state of physical degradation, inhaled these fragrances joyfully. By contrast, when she left us and the stale, sickening odor of burnt human flesh, which covered the camp like a blanket, crept over us again, the atmosphere became even more unbearable.

There was a huge amount of scientific evidence for the hypothesis that each race had its own peculiar odour. This was boosted by numerous personal testaments. Jack Holly, a US marine who led patrols behind the lines in Vietnam, said his life had been saved by his sense of smell: 'You couldn't see a camo bunker if it was right in front of you. But you can't camouflage smell. I smell the North Vietnamese before hearing or seeing them. Their smell was not like ours, not Filipino, not south Vietnamese, either. If I smelled that again I would know it.'

It seemed that we convey a large amount of information in the odours we emit. Doctors have been diagnosing patients by their scent for centuries. Diseases reportedly detectable by odour range from diabetes to insanity. The nose in diagnose emerged more prominently in my mind as I read accounts of dogs and rats being trained to detect the scent of schizophrenia.

Shortly before the library closed I honed in on the history of smell detection by British law enforcement. A reference in a book called *Curiosities of Savage Life*, published by James Greenwood in 1841, mocked the English police force for being 'behind the Lebashi of Abyssinia' because they were yet to learn the secret of 'how to track a thief by the organ of smell':

> The Lebashi (thief catcher) is much feared . . . When a theft has been committed, the sufferer gives information to this official, upon which he sends his servant a certain dose of black meal compounded with milk, on which he makes him smoke tobacco. The servant is thrown into a state of frenzy, in which state he goes from house to house crawling on his hands and feet like one out of his mind. After he has smelt about a number of houses, the Lebashi all the time holding him tight by a cord fastened around the body, he goes at last into a house, lays on its owner's bed and sleeps for some time. His master then arouses him with blows, and he awakes and arrests the owner of the house . . . The person into whose house the entry was made is regarded as the thief, and is forced to pay, whether he be innocent or guilty. No wonder that the population trembles when the Lebashi is seen in the streets, and that everybody tries to be on good terms with him, as there is no saying when he will make his appearance in a house.

It sounded similar to what they were up to nowadays, but using dogs and different procedures. The article I had come across in *Fortune*, during my research on the use of dogs, described how a suspect is placed in a line-up of volunteers. A police officer with a dog on a leash will identify the suspect, supposedly by scent. I tried to imagine how I would feel locked up in a cell on the word of a dog. The article suggested that handcuffs were often left on the real suspect and that the dog might notice these and bark at whoever was wearing them or at whoever they 'sensed' their dog handler wanted them to bark at. I remembered the tale of Clever Hans, the German horse widely believed capable of solving complex mathematical equations at the beginning of the twentieth century. When presented with an equation, Clever Hans would tap the answer with his hoof. His owner took him round the country, demonstrating his abilities free of charge to the public, who were so impressed and astonished that a public commission was set up to investigate Hans's powers. Much to the disappointment of his owner – a mathematician who genuinely believed his horse had been effectively and brilliantly trained – the Director of the Psychological Institute discovered that the horse had no understanding whatsoever of mathematics, but instead reached the correct answer by reading small and involuntary body movements of the questioner.

The *Fortune* article revealed how most comprehensive scent line-up studies suggest that trained dogs are wrong 30 to 40 per cent of the time. Despite this statistic, scent line-ups are admissible in courts in the United States, the Netherlands, Poland and numerous other countries to which we could easily be extradited. The STU 100, the Scent Transfer Unit, vacuums scent on to a gauze pad, which is then placed in a plastic bag, frozen, and stored in a 'scent bank' where it is kept until trial time, when the pad is thawed and offered to a dog as evidence. Scientists fly around the country testifying against it, but it continues to be used.

I read about how generously the research staff at the International Forensic Research Institute at Florida had been funded over the years. The institute had just been awarded $246,634 by the US government 'to design, develop and test a method to improve the performance and scientific defensibility of dog teams used for human scent identification'. 'Scientific defensibility' struck me as an interesting choice of words, presumably reserved for techniques with no scientific foundation.

But they wouldn't be pouring all this money into odour research just to beef up the science behind this form of evidence, would they? I supposed that with Proceeds of Crime legislation it was a good investment. A number of law enforcement agencies have been given rights to retain 50 per cent of confiscated cash for financing research projects directed towards more efficient operations and technology. 'Who can tell the difference between your "innocent" bills and those of a drug trafficker?' asks an article in the *Florida International University Newsletter*. The answer was a dog, if substantiated by Florida University expert witness for 'dog alert cases'. According to the article, this team's evidence 'can convince a judge and jury to send a trafficker behind bars and add hundreds of thousands of dollars in forfeited proceeds to the coffers of federal prosecutors.' I was beginning to see a pattern: the money from these 'coffers' could be used for investing in more science and technology to 'recover' even more money and so the profitable partnership between law enforcement and forensic science expertise would blossom endlessly.

Unsavoury though I found this development, I wasn't satisfied that this represented the extent of governmental designs. We were at war at the moment, I remembered. We were engaged in a War on Terror. Security analysts predicted that this war would last for at least a decade, if not longer. The information I had been given at the MoD conference was unclassified. Presumably a great deal more was going on that was confidential. I realised that it could all be in my head but I couldn't release myself from the sense that something very fishy was going on with this smell business. I didn't have time to read the articles on smell allergies and attempts to ban personal fragrances in public places, and wasn't yet sure how they were relevant. I gathered up the books and joined the painfully long queue at the collection desk. I returned the material and left the library, heading for Soho where I'd arranged to meet a friend from the Department of Trade and Industry.

Taiwan mislays millions of honeybees

The Register: Sci/Tech News for the World
Lester Haines
26th April 2007

Taiwan's beekeepers are reporting the mass disappearance of millions of honeybees, Reuters reports.

According to the country's TVBS television station, around 10 million bees have gone awol in the last two months, with farmers in three regions reporting heavy losses. One beekeeper on the northeast coast told the United Daily News that six million insects had vanished 'for no reason', while another in the south said '80 of his 200 bee boxes had been emptied' . . .

CHAPTER EIGHT

Moths and Madeleines

Large insects were advancing along the black floor, their wings outspread. More could be seen forcing themselves through a small hole in the pane of the conservatory door. More still sailed down from the roof, hurtling blindly forwards in the semi-dark . . . They advanced, a disorderly, driven army.

A. S. Byatt, *Angels and Insects*

I'd met Martin shortly after moving to London, some ten years ago, when I'd been dating a friend of his, and we'd stayed in touch ever since. He'd graduated from Oxford with a first in philosophy, politics and economics and taken a job at the Department of Trade and Industry. Being the government department charged with regulating the export of arms, he'd hoped the job would feed his boyish fascination with tanks and military equipment. Although he rarely divulged the details of his day-to-day tasks, I got the impression that they weren't as compelling as his compulsive collection of books and memorabilia. Though clever, funny and charismatic, few of his eclectic selection of friends shared his interest in the technologies of war and I'd watched him shunned into silence at a number of social gatherings. I told him I wanted to pick his brains on military research, correctly anticipating his willingness to oblige.

He was waiting for me at the bar in the Dog and Duck in Soho. He'd had his hair cut since I'd seen him last and looked like a schoolgirl's pin-up. His dress sense had improved markedly since his marriage to a Norwegian fashion designer. After easing each other into the bar with a few drinks and some idle chit-chat, I told him about the conference I'd been to and about the sniffer bees.

'So I was wondering how I could find out more about why the military is so interested in them.'

'Look into chemical and biological warfare,' he said, and took a swig of his third brandy. 'It's a messy area legally. All needs to be revised. Full of loopholes.' The white of his elbow rested sorely against the warmth of the mahogany bar surface.

'That's interesting,' I said. 'But I hadn't planned on getting into defence issues. It's not my bag.'

'Honestly,' he said, 'the history of biological warfare is fascinating. You should check out the Animals in War exhibition on at the Imperial War Museum.'

'I'm sure it is.' I watched him signal to the barmaid. 'I noticed that they've erected a monument to animals in war by Hyde Park.'

'Did you know,' he took a leather wallet out of his back pocket and removed a £10 note, 'that US scientists devised a plan to drop bat bombs on Japan in the Second World War?' He handed his money over to the barmaid and passed me my drink. 'They even experimented with cats. They wanted to tie them to bombs and then *drop* them from aeroplanes on to German warships. They reasoned,' he said with a broad grin on his face, 'that, as cats hate water they would do anything to get on to the deck of the ship and so they'd *guide* the bomb to its target.'

'Did it work?'

'No!' He laughed. 'The cats became unconscious mid-air.'

'Oh dear.'

'A pigeon-guided missile, however,' he said, crossing his arms and pointing a finger at me, 'would have worked.'

I raised my eyebrows and waited for him to explain.

'Burrhus Skinner designed one. Trained pigeons peck at an image of the target, activating a guidance system that keeps the bomb on the right path until impact. If you think about it, it's not that different to training computers with different algorithms, like they do now.'

'Amazing,' I said. 'It sounds like good fun.'

'Oh no. It's nasty stuff. Unlikely that any serious thought was given to biological warfare until the First World War – though the concept was recognised in medieval times. They used to catapult plague-ridden corpses into besieged cities and throw the bedding of smallpox victims at their enemies but it was all pretty primitive stuff. It wasn't until the First World War, when intelligence started coming in about the use of mustard gas by the Germans, that we started to give it any serious thought.'

'Mustard gas,' I said. 'Smells like garlic.'

Martin paused, a perplexed look clouding his face.

'I missed the presentation but got the hand-outs for the first talk at the conference I went to,' I started to explain clumsily. 'I got given a calendar from the Second World War. Each month has got a picture of a semi-naked girl and the name and smell of a different gas.'

'All soldiers would have been trained to recognise the odour of different gases by the Second World War,' he said. 'In fact they were never used. People were too horrified by the concept after the First World War. Their use in warfare had been banned under the Geneva Protocol in 1925. But everyone was stockpiling them just in case. Porton Down was set up during the First World War to experiment with them. They even have their own animal breeding unit – used to be known as Animal Farm.'

'For biological warfare?'

'Both. Biological and chemical. Both these forms of warfare were prohibited by the Chemical Weapons Convention and Biological Weapons Convention. Because everyone is well aware of how easily these kinds of weapon could get out of control. It's not really talked about.'

'It sounds really interesting, Martin. Fascinating in fact. I hope I get a chance to look into it. I am wondering more about law enforcement applications right now though.'

Martin looked at his watch. His wife would be expecting him home soon.

'Am I right in thinking that the police and military are working closely together these days?' I asked him.

'They have been ever since the end of the Cold War. There are powerful vested interests in maintaining the huge armaments industry that was created in the 1950s. So they've been forced to find civilian applications for military equipment. Which they have. The surveillance industry is predicted to become the fastest growing sector in our economy.'

'That doesn't sound good.'

'No. In fact President Eisenhower gave a speech warning against this sort of development in the 1960s. Interesting thing for a former army general to have said.'

'What, he warned against surveillance?'

'No, about turning the guns on ourselves. About arms companies dictating national policy. Like they've been accused of doing in the EU. One of the two official reasons for the European Commission's decision to fund the development of surveillance technologies is to ensure competitiveness in the global security market for European companies. Arms companies have been badgering the EU about the amount of support American companies get from their government in the wake of 9/11.'

He took his bag off the hook under the bar and stood up to leave.

'Check out the biological and chemical warfare treaties,' he said gravely before downing the remains of his drink. 'It's full of loopholes. In danger of falling apart around them.'

I thanked him for the tip and promised to look into it.

Police dog bites Hazelwood student

News Tribune

Kyle Lowry

31 January 2006

Officials not sure what made dog bite Courtney McGarry

An 11-year-old Hazelwood Middle School student was bitten on the face by a police dog at school Monday afternoon.

Following an in-class demonstration on police activities, sixth-grade student Courtney McGarry stopped to pet Condor – a patrol and drug dog handled by New Albany Police Department Patrolman Mike Isom – on her way out of the classroom. But instead of a nuzzle back, McGarry was bitten on the cheek.

'We're not sure what made the dog spook,' said Dave Rarick, director of communications at New Albany-Floyd County Schools.

CHAPTER NINE

Headspace Defined

It felt odd to be back in the office, after the MoD conference. Ascending the lift shaft with me in a large metal box,
a wispy white-haired woman gesticulated wildly in the air
as if fighting off a pestilent dust and muttered faintly 'Oh
dear . . . oh dear . . . files . . . so many . . . oh dear . . . so
much to file,' before racing through the opening doors
and disappearing in a fluster down the first floor corridor.
Streaks of sunlight painted medieval arches on the walls
and pallid faces of silent trolley pushers weaved their way in
and out of doorways, depositing documents. The building
was filled with the people I imagined Aldous Huxley was
referring to in his essay on *Literature and Science* as having
'only enough intelligence to do what they are told and only

the devotion required to come to work on time'. Harsh words. These wandering souls had probably shone brighter in the past. Their lives were suspended by the routine they had mistaken somewhere along the line for reality. Standing on the landing, I felt I could easily have been in the corridors of a prison or a mental asylum. Everyone groaned over their hatred for the place but no one showed the inclination to escape. I was wary of meeting the same fate. A good friend of mine had recently left. A week or so before his sudden departure I was telling him about my wish to leave. 'You just have to walk out the door,' he said. 'It's that simple.'

Endless bureaucracy had ground any interesting facets of the job into an unpalatable pulp. I applied to go part-time and management told me they would accede more willingly to my request if I could find someone to job-share with. The first colleague I asked declined on the basis that she 'wouldn't know what to do with the free time'. I'd yet to hear if they'd let me go down to three days without a job-share.

My office was the first on the left past the security doors. I held my pass up to the reader, entered the corridor and opened my door. Garth was in before me. He always was. He was the perfect civil servant. He got in at 8.15, ate lunch at his desk while reading the BBC News website, took a twenty-minute walk before resuming work and left at 4.30 on the dot. And he always had an umbrella on him. In the 1990s he had worked as a solicitor for tree protestors at the site for the Newbury bypass, and before that for dance enthusiasts criminalised by their repetitive beats. The work had made him cynical and now that he had a mortgage to pay and a baby to feed, he took to the security of civil service employment like a duck to water. He still retained his spark and his conversation was one of the few stimulants available in the office. He'd studied history and literature at university before training as a lawyer. He was an avid reader and particularly liked 1960s psychedelic literature and science fiction, especially any that dealt with the human/machine interface. He was also interested in Nazi history, news stories about UFOs and secret US military designs for world domination. I did what I could to encourage his outside interests. At first he feared his occasional dives into cyberspace would be detected but he was slowly gaining in confidence and embracing his technological freedom.

'Good morning Garth. How's it going?'

He showed me a picture on his mobile phone. It was of our office door, closed in front of him. He'd taken it from his desk.

'I showed this to my wife yesterday.'

I laughed.

'It's not funny. I'm like the prisoner in *The Man in the Iron Mask*. I never see the outside world and once a day they open my door and bring me my files and that's like my food.'

'That's right,' I said, nestling into my corner at the other end of the room and switching on my computer. While I waited for the machine to come to life I set about making our coffee. Garth had come to look forward to his mid-morning coffee-break. He'd been drinking caffeine-free tea for months in an attempt to improve the health of his heart but the smell of fresh coffee had eventually proven too much for him to resist.

'How are you then? Oh yeah, how was your conference?'

I told him about the research being undertaken, about the people I had met there and about the little old moth woman, or entomologist as she called herself, who had congratulated me in the ladies' for sitting through so many scientific talks despite my legal background.

'There's one technical term that is giving me difficulties as people keep referring to it and I don't know what it means,' I'd told her.

'What's that dear?' she had asked me.

'Headspace.'

I posed the question to Garth. 'What do you think it is?'

'I don't know. In 1960s literature it means space to think, a sort of private mental space. Gimme some headspace man!' – he jutted his head backwards and forwards awkwardly but enthusiastically between his stiff shoulderblades – 'Kinda thing.'

'What do you think it means now?' I said.

He looked at me sceptically. I could only see his head and the top of his arms from where I was sat. It gave him the appearance of a mechanical owl. His pupils crossed over to the computer and he clipped in his arms, typing something into his keyboard.

'Same thing,' he said and read to me from his screen: 'Despite attempts to humanise large open plan office environments, these spaces are often tiring and unproductive for those who inhabit them on a daily basis.'

'Got to agree with the man there,' he added.

This study set out to explore ways of providing open plan office workers with increased levels of privacy through the design of objects used on and around the desk. From these experiences a rationale was developed that directed the project towards the provision of greater psychological privacy (termed "head space") over the erecting of purely physical boundaries.

'That's what it means,' he shrugged, 'psychological privacy.'

I thought back again to what Roger the American free-lance consultant had said about Marcel Proust and about odour being the nightlight in the bedroom of memory. A distant yellow lampshade shone on the corner of my mind. Shadows flitted across a book-laden desk under a small window. I wondered why Roger had mentioned it.

'Where are you reading this from?' I asked.

'Tim Parson's website. He's an artist at the Royal College of Art.'

'Well that isn't the meaning of headspace to scientists,' I said, irritable at having not yet displayed my scientific knowledge. 'It is the technical term for the area surrounding a substance in which its odour can be detected.'

'Scientists,' he said, 'are mad. They are completely divorced from reality. Noam Chomsky described Nazi scientists as "the most brilliant, humane, and highly educated figures of modern civilization, working in isolation, and so entranced by the beauty of the work in which they were engaged that they apparently paid little attention to the consequences".'

I suspected he was reading from his screen again.

'I mean these scientists you met,' he asked, 'has it occurred to them that they are turning Mother Nature into a government spy?'

Our door opened and a tall man deposited a pile of files on Garth's desk. No longer able to see his face, I turned to my computer screen, opened my email box and blitzed my way through multiple requests to 'cascade' information about civil service pension reforms, fire drills and stationery orders. I emailed Ben Teckler to thank him for the conference and tell him how lovely it was to have met him. Ben replied promptly to my email.

'Hi Amber . . . I never got a chance to ask you at the conference but are you on the side of the defence against, or prosecution with, evidence obtained from olfactory surveillance . . . ?'

I was slightly unnerved by the Ministry of Defence's desire to verify what 'side' I was on *after* having met me, but from a legal perspective this was an interesting question for a scientist to ask. I replied that I was 'on the side of the law'. Odour evidence, like any other forensic evidence, should be neutral, and therefore as capable of proving innocence as guilt. I added that I had an open mind about the reliability of odour evidence and felt I should learn more about the science of olfaction before making my mind up about its probative value. I explained that I had a very poor grasp of science and asked his advice on which branch to concentrate in my research on olfaction. I sent him a copy of the New South Wales Ombudsman's report on sniffer dogs that had found them to be wrong 73 per cent of the time and concluded that their deployment led to groundless investigations of innocent members of the public.

'Story that might interest you on the BBC News website today, Amber.' Garth was eating the sandwiches his wife made him.

'What's that?'

'Jake the Labrador is so popular with schoolchildren he gets his own email address.' He laughed as if confident of annoying me.

'I am not interested in all dog stories Garth. Just police sniffer dogs.' This was becoming a problem. Friends, confused by my sudden interest in dogs, had started sending me huge amounts of canine paraphernalia. One suggested I formalise my network of informants and name it 'Dogwatch', an idea I had adopted.

'This *is* a police sniffer dog. Schoolchildren are being encouraged to email Jake the Dog for advice on drugs.'

'For fuck's sake.'

'Surely that's good though isn't it Amber? You wouldn't want children on acid?'

'They might as well be if they are going to email dogs. It's completely barking.'

Garth groaned. He didn't like puns and I was finding them increasingly difficult to avoid. 'I'm off for a wander. See you later.'

I waved at him without looking away from my screen.

A message entitled 'HELLO FROM PARIS' sat in my inbox. It was from Roger. He hoped I'd give him a call if I was ever in Paris and attached an article he thought might be of interest to me. It revealed research showing that homosexual and heterosexual men respond differently to odours believed to be involved in sexual arousal, and that gay men respond in the same way as women. 'The question of whether human pheromones exist has been answered,' commented scientists. 'They do.' As I tried to compute the meaning of this strange communication, a female colleague popped in with cakes and I accepted one guiltily. Garth and I wished our colleagues would stop doing this as we were unwilling to reciprocate, being well aware of the cost of these cakes.

'Any news from Tom?' she asked. I'd ended up telling her the whole story of our disastrous dates one morning over the cakes. She'd found the one about the fish coat particularly amusing. It had taken place a few months after the first date, when I'd passed out in the club toilet where Tom had found me and carried me home. I'd been to stay at the house of my deceased grandparents in Wales where I found a cool 1970s coat in the cupboard. I returned to London with it and put it on when I arrived at the station. I remember thinking my heightened olfactory sensitivity must be the result of a brief countryside respite from the stench of the city and I couldn't get over the pervading smell of fish everywhere. Dinner had been awkward. We'd sat opposite each other over a red and white chequered table-cloth. The only other person in the restaurant was an irritatingly flirtatious Italian waitress. Further details of our night out emerged from his lips, including finding me passed out for a second time in my bathroom at home. I struggled to make attractive conversation in the face of this corrosive image but he'd remained awkwardly distant from me throughout the meal. I realised when I got home that the smell of rotting fish had been emanating from the coat, which I'd had draped on the back of the chair throughout our pizza consumption. I'd texted him that night to apologise for the smell of my coat. On re-reading my text after receiving his attempt at a reassuring reply, I realised I had apologised for my coat being so small. It had seemed that every message I sent him was destined for a tapestry of unattractive madness and I'd resolved after our date to respond to his updates on the work he was doing in a professional manner.

'He's coming round for dinner this evening,' I said. 'Not sure what to cook him and my house is a tip as I have been obsessively researching something and haven't had a chance to clean up.'

'You can't cook for him!' she cried.

'I can't?'

'No! He's been running circles round you for months.'

For all I knew we were just meant to be friends, the kiss on our first date had been a drunken accident, and the fact that he texted me occasionally meant nothing more than that he wanted to stay in touch.

'Just order some take-away pizza and a beer,' my colleague instructed me.

'I can't get him to trek half-way across London to my

studio flat and then order a take-out. He'll think I'm trying to bed him. I've got to cook.'

'Amber, you can't cook for him. Believe me,' she said, flicking her blonde locks and placing the other hand on her hip, 'I know how boys' minds work. You've got to play it cool. Text him now and tell him you've changed your mind.'

I'd been looking forward to this evening all weekend. I was really looking forward to his take on the MoD conference and whether he'd looked into the possibility of bio-surveillance. But I could see her point. I had a lot of dignity to regain in our relationship and maybe cooking wasn't the best way of doing that.

I took out my phone and did what she said.

'Can't be arsed to don apron this evening. Fancy a drink instead?'

'Well done,' she said. 'Let me know what he says.' She winked and stuck her thumb up at me before leaving the room.

Garth came back in a few minutes later and stared at the chocolate éclair on his desk.

'Christ. We've got to start refusing these.'

'I know,' I said absent-mindedly. I was reading up on the latest initiative of the Defense Advanced Research Projects Agency (DARPA) in an attempt to take my mind off the unfolding scenario with Tom. DARPA was a national security agency in the US. It had invented the internet. Primarily concerned with the development of technologies to incapacitate or kill future American enemies, its latest programme was called 'Bio-Revolution' and its aim was to 'harness the insights and power of biology to make US warfighters and their equipment . . . more effective'. In recent years, DARPA had been encouraged – some said for PR purposes, and others economic – to extend its research tentacles into 'dual use' technologies that are valuable for both civilian commerce and national security.

The 'Unique Signature Detection Project', formerly known as the 'Odortype Detection Program', is designed to help identify terrorists based on scents they secrete in their sweat, tears, urine and other bodily fluids. According to unnamed experts, a person's smell is so unique it offers the military an alternative method of identification, as effective as retinal scans and fingerprinting but far less invasive. It can be used to identify and distinguish specific 'high-level-of-interest individuals', said DARPA spokeswoman Jan Walker. If the immune system does in fact create a unique smell that helps identify individuals, explained one of the DARPA-funded researchers, the process could also be reverse-engineered, allowing scientists to track down a specific type of immune system. This would be beneficial for organ transplantation: if a donor's odour-type is similar enough to a recipient's, doctors can be reasonably confident that the latter's body will accept a donated organ. Also, if specific diseases can alter a person's odour, smell detection could prove to be an effective tool for early diagnosis.

The Technical Support Working Group, or TSWG, a US interagency research and development programme for combating terrorism 'at home and abroad', has put out an invitation to tender for the manufacture of a portable device 'for canine handlers to collect human scent for future use to track a specified target'.

An article in the *Observer* referred to a leaked memo from GCHQ, the British intelligence agency, stating that odour identification was being evaluated for possible use in the UK. QinetiQ, the leading British defence and security company that had formerly been part of the Ministry of Defence, confirmed to the *Observer* that it has an expert who deals with the degradation of human bacterial cell culture on the skin. Some security experts told the *Observer* that they anticipated the technology would develop sufficiently to allow police to identify an individual in a large crowd purely on their scent. I sent off a Freedom of Information request for all information obtained by GCHQ on human scent identification systems.

'The thing is,' I said out loud, 'government press releases describe odour detection as a non-invasive method of detection.'

'It probably is non-invasive.' Garth was getting his coat on. He had the afternoon off. 'If they can do it at a distance.'

'Just because it doesn't involve sticking a prod up my arse doesn't mean it isn't invasive,' I said, trying to get his attention. 'It boils down to how we define physical barriers, or legal ones for that matter. What's the point of having rights and rules on taking our fingerprints and DNA if they can get it all from a sniff? The odours I emit can purportedly be used to work out my emotions, my state of health, racial origin, sexual preference and immune type. Some say the way we smell, our chemical aura, is in fact our soul. If it is, then the authorities have no right to examine it.'

Garth looked across at me as if I was mad and reached for his umbrella.

My phone pinged. 'No worries. Not drinking this week. Am exhausted. Catch up another time. I'll cook.'

Fuck. It would look a bit naff to suggest a date now. How long would it be before I saw him again?

'See you tomorrow,' Garth said.

'See you.' I waved and turned my attention back to my computer. An email from management had dropped into my inbox and I opened it with some trepidation.

'Ha ha! Freedom is mine!' I shouted out to Garth as the door closed behind him. He popped his head back round the door.

'They're letting you go part-time?' He'd witnessed the arduous series of meetings my application had entailed.

'Yup! As of next week.'

'Congratulations. What are you going to do with your-self?'

'Find out why governments are so interested in olfaction.'

'In what?'

'In the sense of smell.'

'Oh right. Headspace.' He waved his umbrella in the air at me and left the office.

As he closed the door behind him, I opened up the enormous pink file on my desk, barely held together by a piece of tattered white string. It was an old conviction. The appellant had been fifteen at the time of his confession to attempted murder, some twenty-five years ago. He said he was put under such psychological pressure by the police that he would have admitted to anything to get out of the state of fear they held him in. New linguistic evidence suggested that police had been significantly involved in the wording of his statement, despite testifying on oath that they sat quietly while he wrote his own confession. The case was an interesting one and I worked hard on it late into the evening.

Shortly before eight o'clock, I closed up the file and returned to my emails. Ben had replied to my email. He said it sounded like New South Wales didn't have a very good training programme for their dogs and that neuroscience was the principal area for smell research and recommended anything by Linda Buck.

I looked Linda Buck up on the internet. She and Richard Axel had just been awarded a Nobel Prize for their work on human olfaction. Smell was evidently more important than I had appreciated. Buck outlined the importance of her work in a short and slightly creepy video interview. Chemoreception, or the sense of smell, she explained, was the first sense developed for the survival of organisms. Amongst the parts of the brain targeted by odours are the hypothalamus, the 'major control centre' of the brain, which governs our instinctive behaviour – that which we do automatically and without thought – and the amygdala, which governs our emotional responses. I was intrigued by the amygdala. There was something about the sound of the word.

'In fact,' she drawled in a thick American accent, 'we can use the olfactory system to try to gain information about those circuits . . . including the genes that control those very basic emotional responses and behaviors. That's where we're going next.'

Buck was particularly interested in trying to elucidate 'how subordination may be mediated through smell'. She had started to look at the effects of different odours on animal behaviour and had already seen how the smell of a fox paralyses mice with fear, whereas they try to hide when they scent a skunk. The interviewer wanted to know how we as humans make sense of the olfactory system's translation of scattered molecules into an organised map inside our brains. Richard Axel said that this was a problematic question. The issue of who looks down on these points, as he put it, or who sees the activation, was 'the problem of the ghost in the machine'; and, he remarked, 'philosophers have been addressing this problem now since Plato'.

Linda Buck had her own view on this issue. 'There's nobody looking,' she said. 'It's the way it works.' She paused momentarily. 'It could be that it's just going to take some kind of a conceptual or philosophical step for people to say, that's not a problem, we accept that.' So in her opinion, there was no ghost in the machine; and if there was, we should let it go.

The nightlight flickered.

It was late and time for me to leave. I shut down

my computer, turned out the light and closed the door behind me.

The corridor was deserted. Everyone else had gone home by now and I was alone in the halls of justice. It was eerie at this time of night and I watched my shadow pass before me in the windows on my way out. I stood still for a moment looking into the darkness outside the windows, pausing to think of a reason for this eeriness. I tried to conjure up images of shackled prisoners wailing across the building's arches. But I could hear nothing other than my own body breathing. I swiped out and left the building.

A potion to be taken on trust

Guardian
Tim Radford, Science Editor
2 June 2005

Swiss scientists have realised the snake oil salesman's dream: a potion that increases trust. One whiff of a brain-penetrating hormone called oxytocin, and you would trust him with your wallet, if not your life.

Oxytocin plays a role in the bonding between mother and suckling infant; it helps you feel that you 'know' someone you have met before, and it plays a powerful role in romantic love and desire.

Now, Swiss scientists report in *Nature* today, a few molecules in the nostrils will make you more inclined to trust a business partner.

. . . Paradoxically, Dr Fehr and his colleagues began the experiment because one of them believed that oxytocin signalled trustworthiness, rather than a propensity to trust.

'In Germany we have the saying "it is the decision you make in your belly". It means that your emotions are important. When you see another person and you quickly assess the other person's trustworthiness, this is done in milliseconds, probably, and is not something very conscious.

'It may be oxytocin that is involved. We have shown that it has a causal role to play.'

The research could help in a better

understanding of mental problems such as social phobias and autism. 'Of course, this finding could be misused to induce trusting behaviours that selfish actors subsequently exploit,' he warns.

Quite how oxytocin plays its part is not yet clear. But human society functions on trust, according to Antonio Damasio, of the University of Iowa.

'Some may worry about the prospect that political operators will generously spray the crowd with oxytocin at rallies of their candidates,' he adds in *Nature*.

'The scenario may be rather too close to reality for comfort, but those with such fears should note that current marketing techniques – for political and other products – may well exert their effects through the natural release of molecules such as oxytocin in response to well-crafted stimuli.'

CHAPTER TEN

The Awakening of the Reptilian Brain

Most scientists today would predict that in our future the dazzling searchlight of science will burn off the last fading vapours of what the philosopher Gilbert Ryle termed 'the ghost in the machine', leaving souls as relics of history and wrapping minds firmly inside the nets of causality.

Kathleen Taylor, *Brainwashing: the science of thought control*

Every faculty and perception in her passed the boundary line between insensibility and consciousness, so to speak, at a leap. Without knowing why, she sat up suddenly in the bed, listening for she knew not what. Her head was in a whirl; her heart beat furiously, without any assignable cause. But one trivial event had happened during the interval while she had been asleep. The night-light had gone out . . .

William Wilkie Collins, *The Haunted Hotel*

The amygdala is the 'hub in the wheel of fear'. It takes note of all dangerous stimuli from our experience, possibly including those within the womb, and probably some from ancestral memories stored away in the attics of our minds. It conditions our response units accordingly, ensuring that we react instantaneously to all potential threats. Once provoked, it sets in train a series of motions that take milliseconds to impact on the body and minutes for our cognitive process to rationalise and regain control over. The heartbeat quickens as the stomach muscles contract and nausea sets in. Hairs bristle in a hasty salute to the sound of the heart stirring up the rivers of blood below the skin. This is called the 'fight or flight' response and accounts for those anxious and fearful states that catch us unawares. The amygdala is the primary target in fear conditioning. Once activated, it remembers why and retains the fearful association within its wrinkled clutches.

Consisting of two almond-shaped structures, the amygdala is located in the brain's limbic region. The limbic region used to be known as the smell brain. It is still commonly referred to as the reptilian brain because it is roughly equivalent to the brains of present-day cold-blooded vertebrates. The late twentieth-century scientist and political novelist Arthur Koestler saw it as an 'enemy of freedom'. Its arousal, he believed, would lead to the control of our rational minds 'being taken over by those primitive levels of the hierarchy which the Victorians called "the beast in us".'

In evolution, the limbic region preceded the mammalian areas of the brain that promised to liberate us from the shackles of our environment. Despite the primitiveness of the reptilian brain, it retains an important role in humans, the extent of which is yet to be defined. Its *modus operandi* continues to elude the scientific community. Its functions are now deemed essential to comprehending the mind and scientists are busily provoking it into action by firing odours at it, in a bid to trace its neural pathways.

According to Koestler, Orwell's *Nineteen Eighty-Four*

illustrates the dangers of reducing people to a reptilian state in which they lose their autonomy and have their behaviour dictated by state-sponsored stimuli. 'The films shown by the Ministry of Truth in Orwell's *Nineteen Eighty-Four* aim at regressing the audience to a primitive level, and trigger off orgies of collective hatred.'

Koestler records that the first use of the expression 'the ghost in the machine' was by Professor Gilbert Ryle, 'an Oxford philosopher of strong Behaviourist leanings', a man who treated 'genuine thought as a disease'. In a BBC broadcast, Professor Ryle elaborated his metaphor, switching it to 'the horse in a locomotive'. Koestler railed against these 'derogatory and disconcerting' terms for describing the human psyche; 'in the act of denying the existence of the ghost in the machine – of mind dependent on, but also responsible for, the actions of the body – we incur the risk of turning it into a very nasty, malevolent ghost.' Unlike many technophobes, who use the term 'ghost in the machine' to articulate their fears about machines developing consciousness, Arthur Koestler wasn't worried about robots becoming human. He was worried about humans becoming robotic: 'Machines cannot become like men, but men can become like machines.'

In *Ghost in the Machine*, Koestler provides evidence of the ghost's existence in an attempt to arouse our concern for its welfare. He cites an account by Wilder Penfield, a ground-breaking neurosurgeon, of an experiment he conducted on the exposed brain of a consenting patient. Painless low voltage currents were applied to selected points on the conscious patient's brain. Penfield reports on what happened:

> The neurosurgeon applies an electrode to the motor area of the patient's cerebral cortex causing the opposite hand to move, and when he asks the

patient why he moved the hand, the response is: 'I didn't do it. You made me do it.' . . . It may be said that the patient thinks of himself as having an existence separate from his body.

Once when I warned such a patient of my intention to stimulate the motor area of the cortex, and challenged him to keep his hand from moving when the electrode was applied, he seized it with the other hand and struggled to hold it still. Thus, one hand, under the control of the right hemisphere driven by an electrode, and the other hand, which he controlled through the left hemisphere, were caused to struggle against each other. Behind the 'brain action' of one hemisphere was the patient's mind. Behind the action of the other hemisphere was the electrode.

PART II

CHAPTER ELEVEN

Political Evolution

. . . Suddenly there was a terrible roar all around us and the sky was full of what looked like huge bats, all swooping and screeching and diving around the car, which was going about a hundred miles an hour with the top down to Las Vegas. And a voice was screaming: 'Holy Jesus! What are these goddamn animals?'

Hunter S. Thompson, *Fear and Loathing in Las Vegas*

'You know what you were saying about information gathering and total surveillance?' I asked Tom on arrival at his office. It was Saturday. Frustrated at not having heard back from him about dinner, I'd asked if I could take a look in his library.

'Yeah,' he said, unlocking the door and leading me through a dark corridor.

'Well,' I asked as he unlocked a second door, 'how is all this information useful to a police state?'

'Do you know about the miners' strike in the 1980s?' he asked, turning the lights on and sitting himself at a computer. The wall in front of him was stacked high with labelled files: Guantanamo Bay, Abu Ghraib, European arrest warrants, globalisation of police powers, technologies of control, privacy and a host of other topics.

'A bit.'

'Well, GCHQ's telephone network was used to track the miners. When their cars were stopped on the way to protest sites, the police knew enough about them to target punishment or dissuasion appropriately.'

I hung awkwardly behind him as he sat at the counter that ran along the length and width of the room and switched his computer on.

'The publications on crime detection are up on the top shelf. Sorry they've not been properly filed yet.'

They were what I'd told him I needed to come for.

'Thanks,' I said, looking up at the top shelf.

'You can stand on the desk. It's not a problem.'

It was a small room. I stood on the work counter with my back to him and got stuck in to the files. A research workshop on 'Chemical Sciences and Crime Reduction' held in 2004 sponsored by the Home Office discussed how live animals, plants or insects could provide the basis for 'low logistics burden mobile detectors'. The use of insects had limitations in field deployment due to difficulties with location of the organisms but miniaturised tracking devices could overcome these limitations. 'A greater understanding of how insect sensors work could enable production of biosensors with equivalent sensitivity and more readily field deployable.' The deterrent effect of the black and white markings of skunks was praised by criminologist Graham Farrell for it meant that 'it rarely activates its disagreeable and durable defence mechanisms.' And 'animal scent markings, either as warnings or for identification, use a DNA-referencing system identifiable at the individual level.' Law enforcement had 'only recently come close to mimicking the more sophisticated mechanisms of nature'. Ken Pease provided a summary of scientific developments in crime reduction. 'Smokecloaks' sprayed non-toxic smoke into secure environments when under attack and were likened to 'a squid spraying its ink to thwart predators'. The contention that animals did not die in large numbers in the East Asian tsunami of 2004, because they were able to sense the impending doom, had been described by criminologists as having implications 'for the development of remote-sensing technologies'. Deep in concentration, I felt Tom's arm reaching between my legs. He placed a book on the lower shelf in front of my knees and I felt a spark of sexual desire from him as he did so.

Feeling self-conscious, I climbed down from the counter, clasping a selection of papers in my hand.

'Find any dogs?' he asked. I'd wanted to talk seriously with him about the dangers inherent in funding for scientists being primarily for crime and security applications, but I became defensive when he asked me about dogs.

'It's not just dogs. They've got moths, bees, yeast.'

'Sorry,' he smiled. 'Are you done? I've arranged to meet some friends.'

'Yeah, thanks. I've got some good stuff.'

'Copy it if you want, and then we'll get going.'

I copied the articles I'd found and we left the office.

'So,' he said, 'we should meet up for a drink at some point.'

'Or dinner even?'

'Sure.' He smiled again. 'I'll be in touch.'

'There are police officers outside with little cameras in their hats.' Garth greeted me on arrival at the office on Monday morning.

'Really? Must be awkward when they visit the urinal.'

'Thought you'd *object* to this development.' His Welsh lilt escaped him whenever he said the word 'object', as if his upbringing had taught him to view such a pointless stance with a fond amusement. 'Aren't you *anti* surveillance cameras?'

'Not in police officers' hats. Seems like a sensible place to put them. You expect to be under surveillance when a police officer is looking at you.'

'Where *don't* you like cameras?' he asked, lowering his head and resuming his work position.

'Discreetly designed ones, that I only notice after having hitched up my skirt to pull up a stocking. I wouldn't like there to be one in a lift or in the loo either,' I told him. 'And I wouldn't like to be put in one of those police cells with CCTV monitoring me and the toilet bowl.'

'Think the government is collecting embarrassing images of you do you? To use when you run for the presidency?' he said without looking up from his work.

'They might be,' I replied, looking down at my phone.

I texted Tom. 'Police officers with cameras in their hats patrolling the streets.'

'Convicts a lot of people though, CCTV footage,' said Garth, wandering over to the table in the middle of the room and stamping an envelope with the appropriate markings.

'Ask not what surveillance can do for you, but what surveillance can do you for,' I said. 'The odd thing is that in most of the cases I have looked at here, the CCTV stills are of some bloke sat on a bench, not doing anything.'

'Suspicious.' He dropped the envelope in the out-tray.

'Exactly. It's just used to add an aura of dodginess to the suspect. More behaviour to be explained. I've seen some stills, supposedly used to show paraphernalia, scales or something, on a kitchen table, that include summonses for unpaid electricity bills, and rotting food.'

'That's the digital detail for you. It's circumstantial evidence of a need for money and a druggy lifestyle.' He shrugged, taking a new file from his in-tray. It was a different colour to the one he'd been working on before. They changed the colours in a doomed effort to brighten up the place.

'Sure, that's the impact. I could never understand what the difference was between circumstantial and direct evidence, or why the courts used to be so cautious about accepting it until I found a notebook in my drawer with "Drugs Account Book" written in my handwriting on the front of it. I think I would have had a hard time persuading a jury of my reasons for having it.'

'Why did you have it? You might have a hard job persuading me.' Garth leaned back in his chair. 'Criminality is in your genes after all. The Home Office would have had you tagged a long time ago.'

I explained to Garth that I had a fondness for stationery and often purchase notebooks.

'Not sounding like a good defence so far,' he quipped.

One day I had decided that the time had come for me to get organised and start keeping regular accounts. I wrote 'Account Book' on the front and dutifully noted down my income and expenditure for a good two weeks before lapsing. A few months later I decided to dedicate a notebook to my notes on drug law. I took the notebook out of my drawer, wrote 'Drugs' and crossed a line through 'Account Book'. I didn't get round to writing any notes on drug laws in it. Friends of mine have a sound-system we take round Europe every year. They needed a new van and I went with them to visit a man with one for sale. On seeing the van my friends explained it was no good because it had lots of windows, which would advertise the kit to passers by and make it easy to nick. 'No problem,' said the man with the van. 'I can sort that out. I am a carpenter. I build vans for festivals every year. I can build cabinets inside it that will hide the equipment. Here, have you got a pen and paper?' I rummaged around in my handbag and found my discarded notebook. Inside it he sketched his design for hidden compartments. They bought another van in the end and asked me if I knew anyone in the South of Spain with music venue connections that could help them put on some parties there. I asked my dad, and he provided me with the mobile phone numbers of friends of his living on the Costa del Sol. They were all convicted dope smugglers. When I found this notebook in my drawer, and saw what it was called and that it contained inexplicable figures, sketches of hidden compartments and mobile numbers of convicted dope smugglers, I appreciated the dangers of circumstantial evidence.

'Lucky you weren't being investigated then wasn't it?' Garth said.

'That's the thing. I reckon that the legitimacy of anyone is a matter of how closely you look. The very process of investigation builds up a picture of guilt in the investigator's mind. I guess that is what I have against all this surveillance. Could be used to frame someone very easily.'

'Any stick will serve to beat a dog with, in other words?'

'Quite.' I smiled gratefully.

My heart skipped a beat with the ping of my phone and I grabbed it. It was Tom.

'Have lost my SIM card. Who is this?'

'Amber,' I replied.

'Thought it must be you.'

Strange. Surely people texted him about surveillance developments all the time? Or did I contact him more often than most? Were my news alerts the silly ones? Was I stalking him?

'Did you look at Jake the Dog then?' Garth asked me after lunch.

'Yes. Very concerning. The next generation will be even more passive about dog detection than my own are today.'

'What exactly have you got against these dogs anyway?'

'A lot. They aren't reliable for a start, as I've told you before.'

'You *say* that Amber but it is a *scientific fact* that dogs have super-sensitive noses. Dogs have a nose approximately a hundred thousand to a *million times* more sensitive than a human's. I saw a documentary about wolves which described *and showed* how they are able to track a moose, and' – he pointed his finger at me – 'even verify its physical condition from the tiny,' he made a mincing movement with his fingers, 'microscopic even,' he paused, seemingly amazed by his own knowledge, and withdrew his fingers, 'particles of skin and body fluids that a moose leaves on the ground as it passes. They can do this at several miles distance and several hours later. If they were getting it wrong, then the wolf pack wouldn't last long would it?' He looked at me smugly.

'Bats,' I retorted, 'have an extraordinary sense of hearing. That doesn't mean law enforcement should use them to decipher living-room conversations does it? And the bat-eared fox has colossal ears that apparently work like satellite dishes to pick up the minute sounds of tiny creatures moving. They can even hear a beetle larva burrowing in soil. Why don't we enlist some of them in the global war on crime? They could swoop in on people using target words – a vamped-up version of ECHELON.'

'I think you might be losing the plot,' he said, reaching for his umbrella. 'See you tomorrow. I've got the afternoon off.'

I waved him goodbye.

GCHQ had replied to my FoI request about human scent identification. It thanked me and explained that GCHQ was not a government department for the purposes of the Freedom of Information Act and so they were not obliged to comply with its provisions. I pressed reply and apologised for my ignorance. 'Perhaps you would be willing to answer my request even though you are not obliged to?'

They weren't.

Roger had sent me another article, though, entitled 'This is Your Brain on Chocolate'. It discussed brain scan images of people subjected to the smell of chocolate, showing how it lit up the 'pleasure-anticipation neurons'. The personal aim of the food addiction scientist doing the research was to find out 'if I don't want the buttered popcorn, why is it that every time I go to the movies, I'm a goner?' What was Roger trying to tell me about security research on olfaction?

I decided I'd better visit him in person. One of my closest friends, Cass, would be in Paris and this was a good excuse for seeing her. I hadn't seen her for almost a year as she had been in the US. I'd missed her and wanted to tell her all about the MoD conference. She was the one person I could rely on to view my experience as a dalliance with the Dark Side. Born into a privileged and eccentric family, where she'd learnt to cook, shoot guns and talk the hind leg off a donkey, she flew the nest in pursuit of bohemian dreams. After graduating in anthropology, and having spent time in India learning Tibetan and collecting testimonials from tortured women activists on behalf of human rights interventionists, she settled down with her musician boyfriend in a commune. Besides cooking for money and booking gigs for the commune's musicians, she sought comfort in the writings, music, films and people from the 1960s. In the wake of her break-up with the musician, her life became more nomadic and she now spent the summers at her parents' house, working as a private chef, predominantly for rich people she referred to as 'fancy pants', and the winters reading, writing and translating in Paris. She had a passionate interest in other people's lives and an uncanny ability to engage almost anyone in conversation. In their age of retirement, a number of her parents' friends had come out of the woodwork about their secret service involvement. Cass had been surprised, intrigued and inquisitive.

I rang her up and arranged a visit.

Roaches get a robot pal

New Scientist
16 July 2005

A group of cockroaches have found a friend in a matchbox-sized robot called Insbot.

Developed at the Swiss Federal Institute of Technology in Lausanne, Insbot has learned how to mimic cockroaches' behaviour and interact with a colony of insects. The device was developed to show how artificial systems could interact with animals in future mixed societies . . . In addition to a host of touch sensors that allow it to interact with the roaches, Insbot can secrete chemicals that mimic the pheromones with which they communicate. So accepted has the robot become in roach society that it can even lure the insects from the safe, dark shelters they prefer to a much brighter one.

CHAPTER TWELVE

Birds Flying High

For whom, for what, was that bird singing? No mate, no rival was watching it. What made it sit at the edge of the lonely wood and pour its music into nothingness? He wondered whether after all there was a microphone hidden somewhere near. He and Julia had only spoken in low whispers, and it would not pick up what they had said, but it would pick up on the thrush. Perhaps at the other end of the instrument some small, beetle-like man was listening intently to that.

George Orwell, *Nineteen Eighty-Four*

Cass and I were chatting in her sitting room. The flat had once been the servants' quarters of a luxurious Parisian house. The sitting room windows stretched from the floor to the ceiling overlooking a small courtyard. Very little light penetrated the room, with its wood panelling and low ceilings. When Cass was considering whether to move in, we had joked about male visitors having to get on their knees or lie down the moment they stepped in to the apartment. 'You wouldn't even have to ask them!' she'd marvelled.

A cosy kitchen adjoined the sitting room. Cass had

decorated it with pictures of trees and flowers and a large poster hung on a blue wall above the small dining area in the corner; 'You have a right to remain silent, you are a child of the universe, no less than the trees and the stars, you have a right to be here.'

Cass couldn't get over the conference I'd been to. She was particularly impressed with the Gucci suit, stiletto heels and Dolce & Gabbana perfume combo. I curled up on her sofa, content to lap up her admiration.

'You are under deep cover out there baby, deep cover.'

I decided I liked that idea. I wondered whether this thought had occurred to Tom.

'Thanks Cass. Other people I talk to don't seem to understand the psychological resolve it takes to deal with them.'

'Shit yeah. I did a bit of it in the States. I was working for a pro-choice organisation and they sent me to join a pro-life campaign to find out what their tactics were.'

'No way?'

'Chrrh. Yeah. The worst bit was when we all had to tie ourselves together in a group by interlocking D-locks round each other's necks.'

'Fuck.'

'Yeah. And then we had to sing this religious song.'

'That sounds horrific.'

'The locks thing was a bit scary but the singing bit was kinda nice. You know, the way it can be, when a group of people get together and sing.'

'I . . .' She interrupted me before I could express my admiration and cocked her ear towards the window. I stopped and listened to the loud banging in the court-yard. Its cobbled stones were separated from the street of diplomatic embassies and national government buildings by 12-foot high wooden doors. Cass looked worried.

'What do you think it is?' I asked her.

'I'm worried it's connected to the kids upstairs,' she said, pulling the locks of hair up off her shoulders and dropping her forehead to tie them into a bun. 'These two black kids have moved in and the rest of this building is populated by rich fancy pants Parisians. The dudes dress in caps and baggy trousers and Puffa jackets and they are not making any effort to hide the fact they sell weed. People come in and out at all hours filling the corridors with skunk smoke and discussing different types of dope. The caretaker . . .' She took a deep breath. One of the great things about Cass was her ability to talk at an intriguing pace without inhaling until the subject of her monologue changed. 'She's Colombian and can spot drug dealing a mile off. She says there have been complaints and some of the residents are threatening to call the police if they don't keep the noise down. She's a good woman. She's tried to warn them but she can't lose her job you know. And,' she toked on a spliff, 'I reckon the people banging on the door are people looking for weed. The dudes are great, although they did sell me some sugared skunk the other day, but they are nice boys, but their buyers are idiots. Really indiscrete.'

The banging stopped and was replaced by shouting voices.

'Jesus fuck!' she said, darting over to peep above the linen drapes. 'The police are here and they've got a sniffer dog!' She turned and glared at me boggle-eyed.

I didn't find this fact surprising, having heard her account of the situation in the building. But she was freaking out.

'Does this happen to you a *lot*?' She squashed her spliff, wrapped it in a plastic bag and shoved it under the sink. 'I mean we've never had the police here and I have never seen them with dogs except at the Eurostar terminal.'

I followed her to the window.

'Look at it! It's sniffing everywhere! This is private fucking property. They can't do that. Those buyers were outside. They don't even live here! What if the dog takes them to my door? Can they just march in behind it?' She paced the room, lighting incense and waving it round to distribute the smoke. I was becoming immune to the threat of sniffer dogs.

'Do you think the police heard them banging and came because of it?' I asked her.

'I don't know. It might have been the cops hammering. Maybe they caught some kids with weed and asked them where they'd got it. They've probably been looking for an excuse to get in.'

The police dragged the dog out of the communal bin cupboard in the corner of the courtyard and up the stairs to where the boys lived. Fortunately for them, they weren't in and the police left.

'I can't believe that. I mean that's pretty weird isn't it? The *day* you get here a load of police officers with a dog turn up?'

'I guess so.'

I let her vent her anger and astonishment to a few people on the telephone and lay down on her bed. I picked up her copy of *Intelligence in Nature* by Jeremy Narby. Animal and plant intelligence had taken on a whole new meaning in my mind after the MoD conference. I'd forgotten that the phrase usually referred to the degree to which animal and plant consciousness differed from our own. Of course the extent to which animals are automatons or mechanical in their thinking process is very pertinent to their efficacy as police stooges, I realised. It hadn't been that long ago that the intellectual establishment had denied women conscious minds and now they were talking about robots developing them. If animals worked out that they could use their police work to lock up humans, it might not be long before we were looking at them from the other side of zoo cages.

Narby seemed to be suggesting that Mother Nature was more intelligent than we had given her credit for. I read his account of a conversation with a leading bird scientist who had published an article in the journal *Nature* called 'Birds that Cry Wolf'. The scientist said he had observed that some birds, known to act as sentinels and give alarm calls when they sight predators, frequently used their power to deceive other birds. They would deliberately give false alarm calls to make other birds panic and abandon the food they had flushed out of the trees.

Cass came in to the room.

'Can we go to Liberté for a drink?'

We'd discovered the bar last year and it was now our favourite bar in the world. We knew we would be able to go there in our seventies, and get a dance with a handsome young man. We also knew that as young girls, we were obliged to dance with the older men. Everyone there was determined to have fun and that meant everyone had to join in the vibe.

'I think we should.'

The bar was over-spilling with pirate look-alikes when we arrived. Stepping in was like boarding a boat headed for eternal decadence and it showed on the faces of the regulars. The speakers were booming the melodious lyrics of Nina Simone's 'Feeling Good' as we approached.

Momo, the barman, waved to us from behind a sea of bodies. Cass had befriended him last summer. He'd worked as a lawyer in his native Algeria and then as a security guard in Paris. Now he viewed himself as a social worker, trying to keep a motley crew of refugees from reality content, and only stepping in with his muscle when tempers steered off course. We squeezed our way past the friendly geriatric hips and young loud lesbian lips until we reached him.

'*Ça va*, Cassandra?'

They exchanged kisses and pleasantries and I leant against the bar looking fondly on as Nina lulled those standing into a gentle swing.

And I'm feeling good . . .

The hips of the bodies that crowded round the bar dropped with the baseline. Beer dribbled out of the cold glasses and down our arms as we ducked and dived through the lustful throng and back out on to the terrace. The tables were all occupied and we made our way to a conveniently parked car. Cass sat herself on its bonnet, crossing her legs underneath her and tucking in her skirts. She wasn't one for revealing flesh in public and often made fun of me for doing so. I stood opposite her, resting a foot on the bumper. Sweat was pouring down my waist as quickly as the beer down my throat.

'You know, it's kinda dark the animals they are pickin' to train; bees, dogs, rats – they are all classic phobias – I mean maybe that's why they pick 'em. Didn't that guy at the conference say it was an image thing?'

Dragonfly out in the sun you know what I mean, don't you know?

'Yup.' I was watching the long-haired man Cass had snogged last time we were here. He looked even more haggard than last year and I knew she'd be mortified to see him.

'I mean phobias, they are deeply seated fears in some people. The dad of a friend of mine used to work for psychological operations. And boy, those guys *love* to fuck with heads. It's not all about dropping flyers saying how great their liberating forces are. They're a *lot* more sophisticated than that. Did you read those leaked documents showing that detainees at Guantanamo Bay were examined for psychological weaknesses that could be used in interrogations? And I mean look at what they are doing in Abu Ghraib! When that British lawyer, Clive Stafford Smith, *finally* got permission for those guys to have legal representation, the secret services *pretended* to be lawyers the week before the real lawyers came so that when they *did* come the prisoners' heads were so fucked up they couldn't talk to them straight.'

'Wow. That's dark.'

'Well, yeah, you should check out his book. I mean the CIA was created to spy on foreigners and now it's spying on us back home, all in the name of the War on Terror. Which is ironic. I have to give my fingerprints if I want a friggin' mobile phone. Private companies *sell* information about me to the authorities. They even have a list of government critics to hassle in airports.'

'No?'

'Well, that's what a lot of academics and journalists are saying. And there's this new book out in the States, by Johnathan Moreno, a bioethicist, it's called *Mind Wars*. He says they're coming. You remember all that shit they did with LSD and mind control in the 1950s?'

'Yeah. The MK-ULTRA programme.'

'Well neuroscience wasn't that sophisticated then. It's come a long way since. Have you heard of the Roborat?'

'No.'

'Moreno talks a lot about it. It's got electrodes buried in its stomach out of view. They can control its movements via remote control.'

'What, it's alive?'

'Uh-huh. The people who designed the apparatus say its pleasure centre gets stimulated when it's controlled. Apparently that makes it humane.'

'Christ.'

'And I mean – what if they start using this Abu Ghraib shit in law enforcement? They could make it look humani-fuckingtarian! "An alternative to guns",' she said, imitating a media spokesperson. "'If you know that the suspect is terrified of spiders, you can apprehend him with one. You can carry it in this highly portable and fully security-proof jar, that slots in to this vest pocket!" I mean, fuck.'

'My dad is scared of spiders,' I said pensively. 'He told me this horrible story about when he was in prison. He'd been put in an isolation cell, you know those ones they have underground?'

'Yeah but why, what had he done?'

'Nothing. The Drug Enforcement Agency used to write to the prison with fabricated allegations of planned escapes and that was the prison's automatic reaction.'

'Fuck. So what happened?'

'Well it's pitch black in there. It was the winter in Indiana, which apparently is really hardcore, and the cell floor was covered in water that had seeped in from the surrounding undergrowth. He was on the bed and the only light was this thin stripe running along the gap between the floor and the door. And he is lying there, locked in the cell, with nowhere to go and he sees the legs of a spider in the stripe of light. And then he sees the spider duck its body under the door and into the dark void surrounding him.'

'Urgh.'

'Yeah. And then he had to try to go to sleep.' I sipped at my beer and glanced over at the bar. An Aerosmith frontman look-alike winked at me from behind it.

'A lot of women are scared of cows,' I mused. 'Even I got a bit freaked out in a field once when they all started moving in synchrony around me.' I smiled at the recollection.

'Do they have sniffer cows?' Cass asked in amazement.

'No.' I laughed. 'Not that I am aware of. I did come across a proposal to deploy sniffer plants in the proceedings of a chemical sciences and crime reduction conference. I don't think that is a common phobia. Maybe,' I continued talking as I watched women enticing others away from their male companions to dance, 'they pick animals that have good law enforcement puns: bugs, plants, rats . . .'

When I turned to Cass for help with a pun for the canine I realised her attention had frozen some time ago. .

'Sniffer plants?' she gawped fearfully.

'Yes.' I laughed. 'They change colour when they come into contact with substances of interest. Military scientists have essentially re-engineered the plants' survival instincts.'

'Don't you think that's creepy? I mean, surveillance in cities is one thing. I remember reading a comic-strip in a chick rag when I was in high school. It had all your typical sci-fi dystopian concepts; books had ceased to exist, governments monitored everyone's daily activities, sexual relationships were banned and all that jazz. A young girl and a boy fell in love and decided to flee the regime. And I was like "yes go!" when they ran away to the forest. But they got caught by cameras installed there. That really scared me. No escape from The Man, you know? No escape back to nature.'

'Like the news story we saw about French police hiding behind trees in the forest to catch magic mushroom pickers?'

'Yeah, I mean what the fuck?' she said, bouncing off the bonnet on to the street.

'You won't be amused to hear that sheep have the capability to recognise human faces then?'

'Get outta here!' She laughed and took my glass to refill it at the bar. But I was serious. For all I knew there were plans afoot to install them in Passport Control. Why else would scientists be looking into it? As far as I could tell the police and military were funding most of this sort of scientific research. And a computer wasn't necessarily more intelligent than a sheep.

Algerian *rai* rippled through the bar and the regulars got into full swing. I watched a man with sparkles squinting out of his eyes dance in a leather waistcoat. He looked like he was pumping an imaginary pair of bellows in an attempt to enthuse some ladies to get up from their chairs and join him. Bizarrely, his attempt was successful.

Why hadn't George Orwell thought of global bio-surveillance, I wondered. Maybe he had. His most famous books were about animals taking over the world and surveillance. And there was a scene in *Nineteen Eighty-Four* when Winston and his lover escaped to the forest to have sex and Winston got paranoid that a thrush would report them to a beetle-like man.

Cass was taking her time with my drink and I searched for her red head of hair among the undulating faces and beer bottles. I spotted her hip dancing with a tall man who looked like he'd stepped off the cover of an African funk compilation. I abandoned my chair and tried to join her but was grabbed by the man with the squinting eyes. Before we knew what had hit us we'd consumed twenty beers between us and were toasting new friends with shots of tequila at the bar. There was something about the place that made us want to merge with the merry mess.

Big Brother eyes 'boost honesty'

BBC News
28 June 2006

The feeling of being watched makes people act more honestly, even if the eyes are not real, a study suggests.

A Newcastle University team monitored how much money people put in a canteen 'honesty box' when buying a drink.

They found people put nearly three times as much in when a poster of a pair of eyes was put above the box than when the poster showed flowers.

The brain responds to images of eyes and faces and the poster may have given the feeling of being watched, they say.

Writing in the journal *Biology Letters*, the team says the findings could aid anti-social behaviour initiatives . . .

Processing faces

Dr Melissa Bateson, a behavioural biologist from Newcastle University and the lead author of the study, said: 'It does raise the possibility that you could get people to behave more co-operatively or pro-socially by putting up pictures of eyes . . . It would work particularly in instances where people have to make a choice between whether to behave well or badly.'

She said CCTV or speed cameras might be a possible application.

Professor George Fieldman, an evolutionary psychologist from Buckinghamshire Chilterns University College, said: 'This paper beautifully demonstrates that people behave better when being watched.'

CHAPTER THIRTEEN

PSYOPS

Cass had been to the bakery by the time I got up the next morning and stumbled into the kitchen. I tucked myself on to the bench under the wooden table and smiled at the bread, cheese, marmalade, strawberry tarts and fruit juice. Cass loved food, mainly displaying it and feeding it to other people. She was a highly qualified and passionate chef who'd worked in famous restaurants and millionaires' houses. She bitched and bitched about having to take orders from 'fancy pants', but I suspected she secretly loved working with the finest ingredients. She justified her private chef jobs to her friends with anecdotes of how she'd use $100 tips from the Drug Czar to purchase big bags of weed just feet from his fortressed mansion, 'just for sport'. Cass carried the coffee over from the stove, seemingly unaware of what a fantastic hostess she was. I poured myself a glass of grape juice as she divided the coffee between us.

'Of all the people you met,' she said, passing me a cup, 'Roger sounds like the most interesting.'

I thanked her for the coffee.

'I mean who is he? He said he was a consultant right?'

'Yeah, but,' I took a mouthful of coffee, 'he was down in the brochure as being from the US government, so it's no big secret.'

'Well then why say consultant? Why doesn't he have an official title if he is working for The Man?'

'I don't know, but I liked him. He's the guy I've arranged to meet while I'm here.'

'Oh really? He lives in Paris?'

'I don't know. He's here at the moment though.'

'Well that's cool. So why the big interest in smell? It's kinda creepy don't you think?'

'That's what I want to find out.'

'Well, where are you going to see him? And when? I mean we are leaving for the festival tomorrow, right?'

Last night Cass had told me she couldn't bear to return to the US, or the 'belly of the beast' as she put it, without going to the annual music festival in the Dolomites first. She said she'd shown photos of the naked mud-splattered travelling gypsy children we'd seen there to all her friends back home, and they wanted more. She had succeeded in convincing me I needed to recharge my batteries and soak up the festival's positive vibes. The 'pizza fairies', as we liked to call the travelling Italian bakers, who we presumed came from somewhere within the Alps, set up camp in the same place every year and we tended to spend most of our time there, queuing for pizza. Health and Safety was nowhere to be seen and I'd spotted a small boy rubbing dough playfully on his balls before putting it back into the bowl. The womenfolk wore tattered cloth dresses. The men wore loin-cloths and shovelled pizzas in and out of ovens with giant pitchforks. Cass liked to stand in the unruly queue, clasping my arm and gawping at their naked torsos. She told me her main reason for wanting to see them was her admiration for the mud ovens they built on arrival. But I had my doubts.

'Yes, we are and so I should give him a call now. '

'Right. Use my phone.'

'Thanks.' We adjourned to the sitting room and I took out his card and dialled his number. Cass sat crossed legged at my feet. He answered after three rings.

'Hello Roger, it's Amber.'

'Hey Amber! Are you here for the airshow?'

'No,' I said, taken by surprise, 'what airshow?' Cass raised her left eyebrow.

'Oh. I just thought 'cos you were with the Ministry of Defence you might be here for the airshow that's on in Paris right now.'

'Oh no. I am here on holiday.' I paused. 'And I don't work for the Ministry of Defence.' Cass started giggling and I had to kick her to stop.

'Oh right. So where are you staying?'

'In the seventh arrondissement.' That was going to confuse him. The area was occupied primarily by embassies and the French Department for Defence.

'So do you want to meet for a coffee, maybe this afternoon, or, I mean do you want to go to the airshow tomorrow? Have you got any of your ID stuff with you? It'll be difficult without it 'cos Chirac's gonna be there.'

I explained I was only in Paris until tomorrow morning and asked to meet for coffee instead. He suggested Starbucks, which struck me as the most ridiculous café to go to in Paris.

Cass said they were spreading like bacteria throughout the city. I remembered it was non-smoking and gave that as my reason for wanting to meet elsewhere. We agreed to meet in one of the cafés by the Pompidou centre.

An hour later, I had changed into a knee-length silk summer dress and was sitting at a small round table, overlooking the *fontaine des automates,* sipping water and smoking one of my French cigarettes. I watched a large pair of red lips spout water at a revolving female torso in the fountain.

'Oh boy.' Roger arrived at the table and fingered my packet of Gitanes Blondes Legères. 'You blend right in wherever you are, dontchya?'

'Thank you.' I smiled. 'It's great to see you again.'

'Well, you too.' He crossed his legs and ordered a beer.

'So,' he asked, 'how's work going?'

'Well, unfortunately I have been pretty busy at the office so I haven't had time to do much more research into smell. But I want to. I was hoping you might be able to help me decide what areas to look into next.'

'I've moved on,' he said, removing his shades and letting them drop below his neck. The top buttons of his shirt were undone, revealing a hairy chest. 'I'm researching bio-fuels.'

'Oh?' I said, disappointed with his news.

'Oh it's great stuff. You'll know that in ten years. You'll be using it to power your laptop.'

'So, how come you are doing this now?'

'Well, like I said, I'm freelance. I'm keeping a hand in olfaction but right now I am working for this man who's got a bio-fuel technology company. I am helping him out. It's great stuff. It'll revolutionise the energy industry – and it's a clean resource. I can finally tell my college friend, who is an environmental big shot, what I am doing without getting into an argument.' He laughed.

I lit up another cigarette and he finished his beer.

'Hey,' he said shifting position in his chair, 'what are you doing tonight?'

'I'm not sure.'

'Why don't you come for dinner? My wife and I are going out with the boss. He's a good-looking guy. In his thirties, and *very* rich. I think you'd get along.'

'I'll let you know.' There was no way I was going to be pimped out to his boss, and it sounded like that was what he was suggesting.

'Yeah, think about it. Call me later on this afternoon.'

'So, you haven't got any news for me on the olfactory front?'

'I brought you a list of up-coming conferences,' he said, pulling a small bundle of papers out of a plastic bag.

'Oh great. Thank you.'

'Well, let me know if you come across any promising new technologies if you go to any of them.'

'Sure.'

'And you know about odour preferences being culturally dependent, right?'

'No.' This sounded as tangential as his article on the scent of homosexuals.

'Yeah.' He put his shades back on and placed some money on the table. 'Like many Asians love the smell of fermented fish and most Westerners, apart from the Swedes, hate it.'

'How interesting.'

He paid the bill and I agreed to call him later about dinner. Alone again in the café I ordered myself an espresso and sat mesmerised by the red lips bobbing back and forth hypnotically while Niki de Saint Phalle's multi-coloured serpent twirled in the distance behind it. Following Ron Teckler's tip on neuroscience, I'd started to read Owen Flanagan's *The Science of the Mind* on the Eurostar. It recounted how

Descartes had been inspired by a mini-world of hydraulically controlled robots in the French Royal Gardens of the seventeenth century, to explore whether Galileo's mechanical conception of the physical universe could be extended to human behaviour. The hydraulically controlled robots he'd been inspired by sounded a great deal more sophisticated than the spinning and spitting contraptions I was looking at now.

I waited an hour before calling Roger back to tell him that my hostess had made evening plans for us and I wouldn't be able to meet him and his boss.

'So?' Cass asked me back at her house. She was cooking dinner and had opened a bottle of red wine. 'What did you get?'

'Fuck all. He's into *fuel*. Wanted to take me out to have dinner with his boss. Said he was a good-looking guy and very rich. Only reference he made to smell was to ask me if I appreciated that perception of smell was culturally dependent.'

'Huh?'

'Different cultures have different tastes in smells. I dunno,' I said shrugging and taking a seat at the table. 'But he did give me a list of security conferences coming up.' I held up the printed list.

She poured me a glass of wine as I flipped through it at the kitchen table.

'This thing is sponsored by the Homeland Security Department and the Department of Defense. Its aim is to join policy makers with scientists to develop a framework for the future of security. It's all on surveillance. Check it out,' I said handing it to her. 'Randy Oldensaw, the guy who mentioned terminating suspected terrorists, is chairing one of the sessions.'

'Do you think you could get into this thing?' she asked me after reading it.

'I don't see why not.'

'Well,' she put the list back on the table and brought the food over, 'it's oversubscribed and looks pretty high-powered. You can't just go. It says here that a panel assesses the applications. I think they are looking for scientists.'

'Well I'll give it a shot. Privacy is on the agenda. I do work for the government legal service and I have made a presentation to scientists at the Ministry of Defence,' I reminded her.

Cass shook her head and handed me a fork. 'Really Marjorie?' She called me that when she thought I was getting above myself.

'So,' I asked tucking in to her pasta, 'what did you get up to while I was having coffee?'

'Argh, I went to see Bernard and Virginie. They have just got back from taking ayahuasca in South America. Virginie's eyes were spinning like the snake's in the *Jungle Book* while she told me about the magic of talking with plant spirits.'

'Talking with plant spirits?'

'Yeah, you know, it's a Shamanistic ritual. Permits people to cross-over into the spiritual world and make contact with the spirits of plants and animals. Apparently they can shape-shift on ayahuasca, turn into birds and fly over the town to watch what is going on there.' Cass met my sceptical gaze with a shrug. 'I *know* you don't buy plants having spirits,' she said.

'What makes you so sure?'

'A little bird told me,' she scooped up her skirt playfully and chortled.

'I know a plant registered a response on a lie detector test.' Now it was her turn to look at me in surprise, her mouth full of food. 'They rigged it up and it registered a response as soon as the polygraph expert contemplated burning its leaves. But that doesn't make me think plants have spirits.'

'Or that they can read minds?'

'It makes me think polygraph tests aren't very reliable.'

'You're such a friggin' lawyer sometimes.'

'You know that they are using them to test benefit claimants in the UK?' I persisted.

'Hell, they've been using that shit in the States since the 1950s. Doesn't matter how many times they are shown not to work. Sure it's fucked up. Haven't you come across that President Nixon quote about them?'

'No, what did he say?'

'He said,' Cass tilted her head sideways and adopted a strange facial expression, '"I don't know anything about polygraphs, but I do know that they'll scare the hell out of people".'

We got up early the next morning and took the first train to Venice. We decided to stop over for the night as I hadn't been there before and Cass was keen to share its delights with me. She'd been there a number of times and loved it. She was disappointed when I failed to gasp at the beauty of the canals we crossed on our way to her favourite restaurant. I explained that I had already seen pictures of them in the guidebook.

Further opportunity for architectural appreciation was quickly buried as we set foot on a *vaporetto*; the skies opened and the city was flooded with a torrential downpour. After a few minutes travelling down the canal, the water-bus stopped to let us off and we ran to the restaurant. People were selling umbrellas on the way but it was too late for us. Cass and I were both dressed in linen and were soaked to the skin. The waiters made exclamations of sympathy on our entrance. Luigi, an elderly waiter whom Cass had developed a great fondness for over the years, immediately offered us jumpers from the back room. They were dirty and covered with paint but very warm.

The restaurant was small and cosy with wood-framed windows, illuminated by soft candlelight. There was only one other customer in the restaurant. He was hunched over his meal with his back to the front door, facing the kitchen. Dark and carefully combed hair covered the back of his head.

'You must meet my American friend,' Luigi said to Cass. 'Every year he brings me a bottle of the finest whisky.'

I wasn't in the mood for talking to American tourists. I'd been promised a Venetian experience. Cass, on the other hand, seemed to welcome Luigi's suggestion, perhaps out of frustration with the lack of enthusiasm I'd shown for Venice since we'd arrived. Luigi sat us at a table adjoining the stranger, repeating his comment about the whisky and informing him that Cass and her godmother had been coming to the restaurant for years.

'Good evening,' he said. 'I'm Craig.'

'So what brings you to Venice?' I asked, resigning myself to a polite conversation.

'I am a pilot for Delta. I always visit Luigi,' he said filling his glass with wine, 'when I do a stop-over in Venice. The food here is great.'

'Oh really?' gushed Cass, pleased they shared a fondness for Luigi and the restaurant.

Cass ordered a vegetarian risotto and I asked for the seafood platter.

'So, where are you guys headed?' Craig asked.

'To a music festival in Friuli,' Cass replied and asked him where he was from in the States.

They made more chit-chat and our food arrived. I tuned out in an effort to communicate to Cass that I'd rather talk to her. She refused to get the message. I usually enjoyed Cass's enthusiasm for befriending strangers but this evening it was a disappointment.

'Connecticut . . . oh yeah . . . Long Island . . . ten years . . . can't complain . . . must be tough though . . .'

I concentrated on the seafood, catching tail-ends of their conversation.

'It's like Greenpeace when they got caught by the Russians blocking the whaling boat.'

I looked up when I heard Craig say this. Cass had worked as a chef on the *Rainbow Warrior* and her god-sister was a founding member of Greenpeace France. I had a feeling her new friend was about to express a political opinion that would steer the conversation into choppier waters.

'If you decide to play with the big boys you can't scream like little bitches when you get caught.'

Cass looked more than surprised by his words.

'I agree with you,' I said, hoping Cass would play along. 'It's like protesters against the war in Iraq, complaining when they get arrested.'

'Absolutely,' he said turning to me, fork in hand.

'What?' cried Cass. 'You don't think that Amber.'

'We always argue about this.' I smiled at Craig and tried to shoot a meaningful glance at Cass before continuing. 'I think it's an important war to win and the US is making huge sacrifices for the sake of democracy.'

Cass looked confused and upset, seemingly unable to read me.

'We've lost a thousand soldiers,' Craig concurred.

'How many Iraqis have died?' Cass squealed. Evidently these subjects were too dear to her heart for mind games.

'Who cares?' Craig replied.

'Who cares?' repeated Cass, turning to face me imploringly.

I turned to Craig and shrugged helplessly.

'Look,' he said to her. 'I used to be in the military. I was in PSYOPS.'

Cass raised her eyebrows and kicked me under the table.

'What is that?' I asked.

'Psychological Operations,' said Cass. I remembered her using this phrase in Liberté, but I'd thought it was an expression, not the name of an official body.

Craig nodded, clearly expecting Cass to have heard of it. 'It's a military unit that learns everything there is to know about the target enemy,' he explained to me. He looked down at his plate and pierced the last morsel of meat before looking back up at me. 'Their culture, beliefs, likes, dislikes, strengths, weaknesses, and vulnerabilities. Once you know what motivates your target,' he said, lifting his fork, 'you are ready to begin psychological operations.'

'They call it capturing the hearts and minds of the people,' Cass added while Craig chewed on his steak and I thought about Rudyard Kipling's observation that smells 'are surer than sights or sounds to make your heart-strings crack'.

Luigi came over to clear our tables. He was ready to close up and offered to walk us to the boat when we were ready to leave. We thanked him for his offer and he brought us each a glass of grappa to drink.

'Look.' Craig gulped down the remnants of red liquid from his grease-stained glass and reached for the grappa. 'If you guys were in my unit I would kill a thousand Iraqis to save you two.'

I nodded appreciatively, inwardly horrified at the prospect of being in this man's unit.

'And if I got caught by the Iraqis,' he continued, 'I'd be prepared for the consequences.'

'Hell, yeah! You'd bend over and take it like a man.' Cass grinned in mock admiration. It was the first time she'd smiled since he'd mentioned screaming bitches. I grinned at her and she sent me a conspiratorial twinkle from her eye.

We paid up our bills and Craig waited while we took our jumpers off and Luigi returned them to the kitchen. Seeing no way out of it, we left the restaurant in a foursome. As Craig walked behind me I felt his hand at my waist. The rain had stopped. We traipsed over a bridge, Craig's hand brushing past my waist intermittently. The water below was dark green. Old shuttered buildings loomed closely on either side in the moonlight. The walled and narrow streets twisted and turned with such frequency that I lost all sense of direction. Craig stopped suddenly at a street corner and explained we were no longer going his way. We said our goodbyes and he turned and walked into the night. Luigi insisted on walking us to the boat stop as he explained it was some distance. He led us silently round the street corners in a stone maze that had evidently emerged around us while we had been eating. We reached an opening where a nightlight glittered on the water's edge and stopped to await the boat.

'Wasn't he a monster?' asked Cass. 'I feel like I've just met The Man.'

'I don't think he was bright enough to be strategically important,' I replied.

Suddenly out of the darkness his face appeared under the light-stand behind us. I wasn't looking at Cass but I imagined the blood draining from her face.

'Goodnight, again,' he said and vanished. We boarded the boat silently when it arrived. We felt spooked. Luigi appeared as taken aback as we were and accompanied us on the boat, muttering his surprise at the strangeness of Craig's behaviour.

'Was he following us?' Cass asked as the boat traversed the canal.

It seemed unlikely, but I couldn't think of any other explanation.

This strange encounter cast something of a gloom over our visit and it was a relief to board the train to Friuli.

Most of the people on the train were hippies heading to the festival and after an hour's ride we arrived at the foot of the Italian Alps. We got off excitedly and made our way to the exit when something very strange happened. The crowd clustered, thinned and slowed down. The happy anticipation that had whisked us along evaporated and was replaced with a cloud of panic. It seemed we were now being sucked reluctantly through the exit. We were in fact being channelled through a small doorway where a dog handler was standing with a dog. I was so mesmerised by how close I was to the action that I seemed to slow down and sway my bag under the dog's nose. Cass later confirmed this behaviour on my part. She had more sensibly lifted her bag up in the air when squeezing past. I thought I had passed through successfully but then I felt the dog's nose between my legs and lifting up the back of my dress. Humiliating childhood memories came flooding back and I felt confused about the length of my dress. I was called back and told to wait. Another girl who had been stopped chucked her bag of weed on the floor in front of us and was instantly spotted by an undercover officer with an observation post on the stairs. My legs were shaking. I was surprised by my lack of cool. I had no idea what Italian police were like. My research suddenly seemed less abstract.

Eventually I was directed to take myself and my bag upstairs. I smiled at the officer directing me. I was wearing a small and tight blue cotton dress. The officer returned the smile. Upstairs, on another platform, was a temporary police office. I was asked to wait on a bench inside while they dealt with the other girl behind closed doors. Police personnel came and left, passing me in the waiting room. At every opportunity I dived into my bag to find my small piece of hash, but couldn't. It really was very small. The handler arrived and tied his dog to a post outside. I watched him through the window. He and the undercover officer joined me in the waiting room. It was established that I was English and friendly but didn't speak much Italian. My legs had stopped shaking.

'This is wonderfully interesting for me,' I said in my best Italian.

They looked at me in bemusement.

'Yes,' I continued, 'I am researching sniffer dogs.'

'Are you a vet?' asked the handler.

'No, I am a lawyer.'

They laughed at me for being a lawyer but they seemed to like me. The dog handler was quite cute and smiling at me lots. His hat had a cartoon of a dog on the side of it. He had had the dog for eight years. Did I have any cannabis on me? It was possible I had a very small amount in my handbag. He told me not to worry; it would only be an administrative matter. No, I couldn't have a photo with the dog or one of me wearing the handler's hat; it was against regulations.

I found my cannabis and gave it to them. They insisted on searching through every item in my bag.

'There are no other drugs in my belongings. Ask the dog.'

They refused to get the dog to sniff my bags and I was in there for an hour as they inspected my vitamin capsules individually, while an aggressive female officer snapped the fingers of her plastic gloves disapprovingly at me.

CHAPTER FOURTEEN

The Sound of the Beast

The stench in our carriage became very bad indeed. Above our heads we heard scratching and pecking, and after a while we found that a swarm of big birds were perching on the roofs of all the trucks; they had probably been attracted by the smell of our train. We watched them circle and sail amongst the rocks. Now and then one of them clung for a short time to our ventilation gates, pecking with its hard beak through the gaps and flapping with its wings. I had never seen such birds before; they had bald, cadaverous heads and long wrinkled necks like a plucked hen's. We tried to slay them with whatever we had in our pockets, but they always got away.

Arthur Koestler, *Arrival and Departure*

I'd been unable to enjoy myself at the festival as much as I had in previous years. While Cass attended African drumming lessons I made the mistake of reading up on biological and chemical warfare. I found there was a gaping hole in the middle of both conventions. The Chemical Weapons Convention prohibits the use of toxic chemicals as weapons but leaves open the possibility of using them for the purposes of 'law enforcement including domestic riot control'. The Biological Weapons Convention prohibits the use of biological agents as weapons unless for 'peaceful purposes'. The ambiguity of this phrase is being used to legitimise their future development, in the name of law enforcement. Arms control organisations fear that these interpretations will undermine the strict controls on chemical and biological weapons and lead to their use in warfare. I was disturbed by the idea of them actually being used by law enforcement.

It seemed to me, as I traipsed round the grounds of the nature reserve in which the festival was held, that looking through the loopholes had aroused a dormant part of my mind and my thoughts rattled around uneasily in their new surroundings. Even the festival fairies seemed to be feeling the fear this summer. Huge screens projected images of the London bombings in the chill-out tents and a new security firm had been brought in to harass the hippies. Everyone was bitching about them – even 'Dr Zen' who ran the meditation zone. 'If I want things to get worse here I'll call them in.' They rode three-wheeled motor bikes and carried large walkie-talkies. Whereas last year the only drug consumed was cannabis, this year cocaine was being peddled up and down the campsite. Even the beautifully hand-painted signs, 'If you think you are too small to make a difference, try sleeping with a mosquito', filled me with apprehension. There was a fine line between dancehall enthusiasm and violence and the multitude seemed to me to be struggling to maintain it as the nights drew in. A stolen cop-car light flashed in the darkness as it swung round and round amongst dancers humping the floor to 'the sound of the police – ooh ooh – it's the sound of the beast.'

Back in the UK the news was increasingly disturbing. Bombs were quickly followed with hooded police raids and the trigger-happy shooting of an innocent civilian.

More and more sniffer dogs littered the newspapers

with celebrated deployments detecting the scent of drugs, banknotes, explosives, mobile phones, human bodies and DVDs. Little mention was made of the dogs' inability to determine the quantity of banknotes or to distinguish between counterfeit DVDs and legitimate ones. Dogs were good. Dogs had 25 times more smell receptor cells than humans. Dogs saved lives, located dangerous drugs, won trophies and earned more than their police handlers. Sometimes they mauled small children by accident.

In the office my in-tray was piled comically high. At least Garth seemed pleased to see me. His eyes scuttled over to the cafetière as I entered and settled on me as I switched on the kettle.

'How's it been?' I asked.

'Oh much the same you know. A few bombs went off last week, seems we are liable to be shot for running for tubes, even if we're not running, but that's about it.'

'I know. I had to reply to about a hundred texts asking me if I was alive.'

'Where were you then?'

'Trying to get away from it all in Italy.'

'Enjoying yourself then? Selfish of you – don't you know that there is only so much happiness in the world and others need it more than you?'

'So,' he asked, clasping his hands together after I failed to respond to his admonishment, 'any closer to finding out what the headspace invaders are looking for?'

'Have you heard of Jacobson's organ?' I asked him, pouring coffee into the glass pot. It was difficult to know where else to start and I thought its mysterious disappearance might pique his interest.

'Nope.'

'It was discovered and described in humans a century

ago, but has since mysteriously disappeared from the text-books.'

'Where is it?' he asked, leaning forward attentively.

'The external evidence,' I said, pouring the water over the coffee granules, 'consists of a pair of tiny pits, one on either side of the nasal septum, a centimetre and a half above every human nostril.'

'What's it for?' Garth held his fingers up to his nostrils.

'According to Lyall Watson, a biologist . . .'

'I've heard of Lyall Watson,' he said impatiently. 'He is very famous – got more degrees than we have files. Trained under Desmond Morris. Written a lot of books on zoology. *Supernature* is his most famous. Bit discredited though by the hundredth monkey thing.'

'The what?'

'The idea that a learned behaviour spreads instant-aneously from one group of monkeys to all related monkeys once a critical number is reached.'

'Right, well, anyway, according to him, Jacobson's organ opens up a channel separate from the main olfac-tory system to feed an older, more primal area of the brain that monitors airborne hormones and other undercover patterns of information. It affects our emotional awareness and behaviour. He says this system could be the mecha-nism for operating a sixth sense, one that accounts for our ability to detect messages that haven't been commu-nicated in the traditional or obvious way. In fact, he has evidence that the cause of schizophrenia lies in the ability of schizophrenics to detect these chemical signals. Schizo-phrenics perceive these messages and attempt to respond in this non-verbal language. Their confusion, anger and fear comes from the resulting miscommunication.'

'Sounds to me like you are getting pretty close to the US government's plans with this stuff.'

I laughed. 'What do you mean?'

'Well, this sixth sense business.'

'Come off it.'

'I'm serious,' he said. 'The US government has been trying to read and control minds for decades. Remember all those experiments they did with LSD?'

'Yes, the British MoD did them too.'

'Pah! The MoD 'fessed up to some ridiculous experiments with psychic powers recently. It was on the BBC News website the other day. Got a bunch of reputed psychics together, blind-folded them and asked them if they could tell them the contents of some envelopes, which they couldn't. But the US secret services are particularly obsessed with psychic powers,' he said forcefully.

Garth's telephone rang and I eventually got round to turning my computer on. It wouldn't have looked good if management had walked in at 11 a.m. and I'd been sitting there looking at a dead screen. My computer came to life with rabid enthusiasm and bombarded me with contradictory messages to leave and remain in the building on the day of the bombings.

My formal application to attend the science and security conference had been accepted by a panel of international customs officials who now wished me to present a poster at it. Was I happy to do this? I said yes, without the faintest idea of what a poster presentation was and emailed Ben Teckler at the MoD to ask him.

I was going to have to work bloody hard to clear my in-tray in time to attend this thing. The days rapidly merged into mush as I struggled to get everything done.

Ben kindly explained that a poster presentation was a combination of text and graphics displayed on a poster and normally laminated and erected on a board. He attached a copy of the olfaction department's 'From a Whiff to a Wag: A Search Dog Nose Best!'

I spent my last three days in the office frantically and surreptitiously conceptualising the relationship between odour detection and privacy on a diagram, collating relevant images and quotes and wrestling with PowerPoint before working out where the fuck to print something five feet long and how the hell to get it laminated without management working out what I was spending my time on and without bankrupting myself in the process. The conference itself cost £1000. Fortunately the Honourable Society of Gray's Inn awarded me some scholarship money towards it. This, I hoped, might impress the Americans.

I saw Tom walking down the pavement on my way home. I waved gingerly from my bicycle. His face broadened into a happy grin and I climbed off my bike and crossed over the street to greet him with a kiss on each cheek. As always, the ritual blinded me, reducing me to auto-pilot mode. I'd have liked to linger longer, and learned to plant my kisses more fruitfully, but the rush of emotions on coming face-to-face with him blew away my controls.

'Do you want to grab a coffee?' he asked, the moment having passed.

'Yes,' I said, as the street shops came back into view. He swept me down the road alongside him until we reached a table. I sat down and he went to fetch us drinks.

'So what do you make of the Menezes shooting?' I asked, placing my coffee cup shakily on the table.

'It's outrageous. Unfortunately it'll probably be used to fuel calls for so-called less-than-lethal weapons.'

'What exactly are less-than-lethal weapons?'

'It's a bit of a misnomer. Less-than-lethal weapons are supposed to be weapons designed to cause damage without resulting in death. Tasers are less-than-lethal weapons. Only, like knives, most less-than-lethal weapons result in death if used to excess but they're based on a scientific approach and are more sophisticated. So they get advertised as benign.'

'Oh – I meant to ask you. Have you heard of tera-hertz?'

'Oh yeah, the people zapper! Directed energy weapons?'

'Yeah.' That sounded like the device Randy Oldensaw had described. 'But I think terahertz are the waves that they use to penetrate clothing in people scanners as well?'

'Yeah. It's like infra-red. My mate saw some police officers wearing night vision goggles in Tarifa, on the southern coast of Spain. Looking for Moroccan immigrants swimming in to shore. They use infra-red. I wonder if you can turn up the controls when you want to zap people? Detect and debilitate! All in one – military technology for law enforcement. It'll be all the rage!'

'Punishment without law. An interesting idea. It could all be automated. Is that what you are looking into now?'

'When I've got time, yeah. There isn't much in the way of international guidance on their use. It boils down to what's deemed to be "reasonable force". Unbelievable some of the stuff they are coming up with. The technology is described as a "force enabler". It's outrageous. But I'm working on the handing over of our personal data to the Americans at the moment. In fact,' he grimaced, 'I'd better get going. I've got a meeting about it tonight.'

'Of course.' I hastily swallowed the remains of my coffee. 'I'm going to a security conference in Switzerland next week.'

'Well let's catch up when you get back. You can tell me all about it. And I'll send you some stuff on the people zapper if you want.'

'Sure.'

At work the next morning, various committee members had sent me through details of the security conference. It was a week-long affair, in the small village of the Les Diablerets, in the Swiss Alps. Accommodation at a four-star hotel and three meals a day were included in the conference price.

Cass had emailed me an article from *Slate* magazine, by Harvey Rischikof and Michael Schrage, entitled 'Brave New World: Technology vs. Torture: Psychopharmaceuticals and brain imaging could make prisoner interrogation more humane. Should we use them?' She thought the article might shed some light on the types of detection and interrogation technologies the military was looking for.

It started with a paragraph bemoaning the bad world press America received when the photos of Private Lynndie England standing over piles of naked detainees at Abu Ghraib appeared. The fact that the officers had been court-martialled wouldn't, in the opinion of the author, solve the problem America now faced: 'They'll likely claim a Nuremberg defense and argue they were "just following orders",' and this, he feared, would only bring the US further embarrassment. What America had to do now, he said, was to consider 'different ways to procure valuable information'. 'The good news,' the authors said, is that new means 'exist, or will soon.' These new methods, including brain-scans and truth serums, would 'render moot debates about the excesses of Abu Ghraib-style treatment'. They might even be more accurate because they would reveal when a person was lying, the authors suggested, thereby protecting any innocents from making false confessions. 'The outrage,' the authors explained, 'attending the news about Abu Ghraib probably wouldn't have arisen if the images featured detainees who weren't naked, hooded, or sexually posed as preludes to hostile interrogation.' They obviously didn't think the hungry presence of dogs in the photos was to blame for the controversy and, sadly, he was probably right. And most optimistically of all, he observed, 'it's not clear that these minimally invasive interrogation options would cross a hallowed legal line.' The goal of such methods, he assured the reader, was not to update the CIA's notorious MK-ULTRA 'mind control' experiments of the 1950s, which administered LSD and performed other experiments on unwitting prisoners, but 'rather, the point would be to declare that, just as America's armed forces use precision-guided munitions and "smart bombs" to minimize civilian casualties, America's interrogation methods rely upon new technologies to decrease the risk of illegal abuse.' He sounded like he was proposing the terms 'new technologies', 'harm minimization' and 'decreased risk' as a PR package. 'Even if torture and abuse were effective interrogation tactics, they intrinsically undermine the values American society says it stands for. By contrast, using minimally invasive technologies explicitly designed *not* to be harmful represents values that can be defended both at home and abroad.'

Reassuring stuff. Thanks Cass.

I dashed out of the office at lunchtime and bought the *Rough Guide* to Switzerland.

Les Diablerets, where the conference was located, translated into 'the Abode of the Devil'. It was described as a crossroads of the underworld, a meeting place for witches' sabbaths, 'where the damned and all evil spirits gather'. Tales abounded of 'lost souls seen at night, drifting with lanterns alone or in groups though the woods . . . the meadows and hills all around are said to be inhabited by elves, goblins and a local brand of imp.' Trust the dog to have led me to some ghost-filled Underworld.

CHAPTER FIFTEEN

Abode of the Devil

'I- – I'm a little girl,' said Alice, rather doubtfully, as she remembered the number of changes she had gone through that day. 'A likely story indeed!' said the Pigeon, in a tone of the deepest contempt. 'I've seen a good many little girls in my time, but never one with such a neck as that! No, no; you're a serpent; and there's no use denying it. I suppose you'll be telling me next that you've never tasted an egg!' 'I have tasted eggs, certainly,' said Alice, who was a very truthful child; 'but little girls eat eggs quite as much as serpents do, you know.' 'I don't believe it,' said the Pigeon; 'but if they do, why then they're kind of a serpent, that's all I can say.'

Lewis Carroll, *Alice in Wonderland*

For it must be remembered that what scientific men mean by truth is, in the last resort, convenience. Scientific men are pragmatists in practice, whatever they may think they are in theory.

J. W. N. Sullivan, *The Limitations of Science*

On the eve of the conference I was pre-menstrual. A sense of inadequate psychological preparation was no doubt inevitable. Confused indecisiveness hastened me away from my plans of sleep and restoration, into the company of men and alcohol, and kept me there until closing time when I slunk home. I arrived at midnight, opened the door to my bedroom and surveyed the scene. My belongings were packed in a small black trolley-bag and a suit was draped across the wooden rocking chair. I set the alarm for 2.30 a.m., undressed, turned out the light and fell into bed.

A few hours and a dimly-lit coach ride later, Luton airport was swarming with travellers under bright neon lights. I was sleepy and deeply regretting not having had more rest. I had read in accounts of interrogation techniques that sleep deprivation could impair brain function, visibly affecting a person's behaviour even leading on occasion to hallucinations and paranoid delusions. I was already afraid of losing the plot in the strange company I'd be keeping. I found securitised conversation draining and I had a week of it in store. I had yet to receive confirmation that I had been allocated a room of my own. I had no idea who I might have to share a bedroom with if none became available, and feared it was unlikely to be one of my kind. The security services and law enforcement agencies of several continents would be represented at this conference, as well as several arms companies and other security industry hawks. I'd been surprised by the presence of arms companies on the list of conference attendees and then I'd remembered what my old friend Martin had told me about the emergence of 'a security-industrial complex', foreseen by Eisenhower in the 1960s, in which the boundaries between internal and external security, policing and military operations are eroded and the arms companies dictate government policy on the control of national citizens. I now perceived it on the horizon, an ominous warship coming in to dock.

There had been a number of unusual and dangerous landslides in the regions neighbouring the part of the Swiss Alps to which I was headed. The altitude was an additional source of anxiety; the conference brochure stating that we would be at risk of suffering from high altitude sickness and advising me to pack aspirin. A friend had given me some coca-leaf tea-bags to alleviate any dizziness and nausea. As I checked in my luggage, I looked forward to a couple of hours of restorative sleep on the plane.

On the shuttle-bus, I sank into my seat and leaned my head against the nearest pole. My eyes started to water. As my head slid down the pole I glimpsed Professor Hall from the Home Office under the elbow of the man next to me. I hadn't expected other conference attendees to be flying on easyJet and didn't feel mentally prepared to interact with him. I berated myself again for the wasted hours in the pub. As I stood up to face the bus doors, Bee Girl's blonde locks fuzzed my vision. She stood erect at the door of the bus in a black trouser suit, vanilla top and high heels. She looked good. I felt scruffy and unattractive in comparison.

We arrived at the plane and I decided to call out to her as she stepped off the bus. We had got on well at the MoD conference and this was a good opportunity to make her a friend.

She didn't hear me and I called her name again as I stumbled off the bus. She stopped and hovered in front of me without recognition in her face.

'I met you at the MoD conference in November,' I explained.

'Oh yes,' she smiled. 'My, you have a good memory for names. You are the lawyer aren't you?'

'Yes. Amber.' I followed her up the steps to the plane.

'How are you?' She paused at the top of the stairs.

'Exhausted,' I said, sidestepping her into the aircraft. 'I've only had two hours' sleep.'

'I thought I was bad on four,' she said, getting on to the plane behind me.

We hovered together awkwardly halfway down the aisle.

'Do you want company?' she asked.

'Yes.' Sleep would have to wait.

I took the window seat and she settled in the seat next to me.

'You know your friend is on the plane,' I said, strapping myself into the seat-belt.

She frowned.

'Professor Hall.'

'Is he?'

She looked at me wide eyed and gulped. I gestured to the seats behind us. Bee Girl swivelled the upper part of her small body round and buried her face between the chairs. She spotted him two rows behind and I turned my head to look at him. His eyelids rose and fell in rapid succession across his pink eyes.

'I think he's going to sleep, thank God,' Bee Girl said, sitting back down and searching for her seat-belt.

'I thought he was a friend?' I said.

'He's a work colleague.' She fastened her belt.

At the end of the aisle a hostess was holding up a life jacket.

'I made a presentation at a bee keepers' conference recently,' she told me. 'Now *they* are a strange bunch!'

'Stranger than the dog people?'

'I don't know,' she giggled.

'But I thought bees were your thing?'

'Gawd no. I don't really know much about bees. I didn't know *anything* about them until I started this job. It can be very embarrassing because people ask me lots of questions about bee life and I don't know the answers.'

Luton airport disappeared out of the window.

'Are you making a presentation?' she asked.

'Not an oral one – just a poster.'

'On?'

'Privacy.'

'I think bees are probably better than dogs on that front.' She took the inflight magazine out of her seat pocket. 'Don't you?'

'Perhaps. Why do you think they are?'

'Well because you can put them in a box. That way the person won't know that they are being sniffed.'

She clearly hadn't grasped my concept of privacy but she was a good saleswoman and the idea that police could be secretly passing boxes of bees over unsuspecting travellers was intriguing. What reason would the police give for suspecting travellers without mentioning the invertebrate intelligence? Were the police referring to cold-blooded arthropods when they cited intelligence sources? I didn't want to start lecturing her on privacy and instead accompanied her round this corner in the conversation. 'They *could* put dogs in boxes I suppose, but that would be impractical wouldn't it?'

She laughed and flicked open the magazine. I took out my guidebook.

'Wow, you are organised,' she said.

'Not really,' I assured her.

We both began to read. I thought about telling her what I read in the guidebook about the fairies and goblins and decided against it. I spent the remainder of the flight in apprehensive silence, my mind conjuring up images of the fearful beasts described in H. G. Wells's *The Island of Dr Moreau*, walking towards me 'with quivering nostrils and glittering eyes'.

We put our reading material down as we prepared for landing.

'Jeremy from the MoD has asked me to wait for him at the airport. He and his team are flying British Airways and arriving about forty-five minutes after us, if you fancy waiting with me?'

'Yeah, I could do.' I wasn't sure who Jeremy was.

'Don't worry, you don't have to. I am just trying to avoid waiting on my own at the airport!'

I would barely have enough time for a nap before dinner as it was and couldn't decide what to do. It would be unfriendly to head off on my own, I might get lost, and it might be interesting to join the gathering in the airport, but I really needed some sleep.

I caught sight of Professor Hall wiping his mouth and rubbing his eyes. He looked more human now. He'd probably slept well and I was envious. He spotted Bee Girl and as soon as the seat-belt sign was switched off he grabbed his briefcase from the overhead luggage compartment, zipped up his anorak and bounded over to her.

'I didn't spot you,' he said before catching his breath.

'Hi Doug. Do you remember Amber?'

He looked at me blankly and blinked.

'She gave a talk at the MoD conference.'

'Oh yes, of course. Hello. Fancy a coffee at the airport?'

'Sure.'

Inside the airport, their luggage arrived first and I was pleased to see them waiting for me to get mine. I'd made two friends. We headed out to the café together. Doug bought us both drinks and sat down.

'I told Alfred I'd meet him here. He ah . . .' Doug ducked his head to the right towards the table as if to indicate this was a bit of a secret, 'works for MI5.'

He flipped his head back into a casual position and glanced towards the arrivals gate.

'I told Jeremy I'd wait for him here,' said Bee Girl, neglecting to comment on Doug's indiscretion. 'I guess they are all on the BA flight. It should be here any minute.'

Feeling faint, I went to the loo and texted Cass.

'Deliriously tired. Bee Girl and Buzz Man on plane. MoD and MI5 arriving to meet us. Am scared.'

'Show no fear agent amber.'

She was right. Only the other day I had read about a mouse genetically engineered by Japanese scientists not to fear cats, simply by tweaking its sense of smell. As a result of failing to associate the smell of the cat with danger, which normal mice are innately wired to do, the mouse failed to flee the first cat it detected, approaching it in a friendly fashion. Apparently this behaviour so surprised the cat that instead of attempting to eat the mouse, it returned a measure of affection. In calling me an agent she was referring to our intoxicated fantasies of Dogwatch as a secret intelligence-gathering organisation, reminding me that this was, of course, a laugh. I was in Switzerland meeting the MoD and MI5 for coffee. It beat the usual day in the office. So what if I was tired? I'd have a fag and another coffee and I'd wait with them till the others arrived.

Bee Girl was listening intently to Doug as I approached the table, smiling broadly. 'I mean I remember when they told me what they were planning on doing I said "Are you sure this is going to work?"' He wiped the sweat from his forehead.

I sat back down at the table.

'Actually the force that tracked down and shot the Brazilian wasn't using this equipment. But still.'

I shyly removed my rolling tobacco, careful to hide what I then noticed was a torn Rizla packet.

'It's a problem with any equipment that works by detecting radiation, the problem you have is that tobacco radiates!' He grabbed my tobacco pouch.

I decided against having a cigarette and took out a banana. 'And bananas –' he said shrugging his shoulders and shaking his head, 'they give off loads of radiation.'

A man in his fifties with a youthful face and short auburn hair appeared at Bee Girl's side, swung his stylishly pink bag round and under the table, leant over and kissed her on the cheek. His eyes were friendly but when he turned to face me I felt the pierce of their steel glints.

'Teckler said you'd be coming.'

Now I recognised him. He was Jeremy Thomas, Ben Teckler's senior in the Ministry of Defence. Teckler told me he'd been furious when he had showed him the New South Wales study on police dogs claiming the dogs were inaccurate 73 per cent of the time and had demanded to know where he had got it from. Teckler told him I had sent it to him.

'Hello again,' I smiled.

Another, younger-looking man lingered a few feet away in the background. His glasses were large, thick and geeky. He didn't greet anyone and looked a little awkward. Behind him was a chubby man with dark hair and a friendly face. They were Jeremy's boys at the MoD.

'Have you seen Alfred?' Doug said as a grinning face and head of ginger hair appeared next to him. It was Alfred. He was in his sixties and his face was dotted with freckles. So this was the man from the Five. He looked, with orange sideburns coating the contours of his face, more like the Cheshire Cat from *Alice in Wonderland* than James Bond.

The pack assembled, we made our way to the station. Inside I copied Bee Girl and purchased a first-class ticket. My bank account would now be empty until pay-day.

On the platform, waiting for our train, the man in the thick glasses suddenly disappeared.

'Where's Johnny?' asked Jeremy. He spoke with the air of a cub-scout leader.

'He forgot to pick his poster off the luggage belt.' The chubby bloke laughed. 'He's gone to try and get it.'

'What?' said Jeremy, momentarily perturbed. He looked at his watch. Our train wasn't due for another twenty minutes. He smiled.

'That is ridiculous. He won't get past security.'

We waited together for the train. Alfred ambled up and down the platform smoking a cigarette.

Ten minutes later, Johnny came running down the stairs on to the platform, carrying a three-foot-long cardboard tube. The assembled group laughed. Within minutes, one of boys, with an approving nod from Jeremy, had taken the tube from its resting placing between Johnny's legs and hidden it. Johnny had forgotten about it again by the time the train arrived.

Johnny and the chubby bloke sat in the second-class carriage. Bee Girl, Alfred, Doug, Jeremy and I boarded the first class. Doug and Jeremy sat at one table. Bee Girl and I sat opposite each other at the table behind theirs. The train left the platform and the Swiss countryside emerged behind Alfred who was leaning cross-legged against the window at the table next to ours. He smiled at us through a cloud of smoke.

'So you work with sniffer bees do you?' he hummed at Bee Girl.

'Yes.'

'And how is it going?' The words grated coarsely in his larynx giving them a strangely gruff edge.

'Pretty well, I think.'

He uncrossed his legs and leaned towards her.

'Now, I think you are really on to something with these

bees.' A squeaky undertone rose up to reverberate around the rough edges of his words, and the word 'bee' seemed to come alive and buzz at the end of his tongue. 'I think they have a lot of potential.'

He followed my gaze and looked out the window behind him at the lake, before turning his attention back to Bee Girl.

'We have been doing studies recently on chemical traces of prohibited substances in lakes. We might be expanding in this area . . .' he took a last drag on his cigarette before extinguishing it in the ashtray on the windowsill, 'checking people's water pipes for traces of drugs,' he said, exhaling smoke. 'That sort of thing. There might be a place for your bees in this sort of operation.'

'Hmmn,' Bee Girl crossed her arms and tipped her feet up at the ceiling. 'I don't think we have looked into their ability to detect odours in water.'

I didn't even know they could swim. And rats seemed a more suitable choice for sewer espionage.

'Well, it might be worth looking into. You know that the director of MI5 is a beekeeper?'

I winked at Bee Girl. I had reason to think this was how animals made it into the forces. A powerfully placed animal enthusiast could cause havoc in security circles. I had read in Charles Pomery's book *State Secrets* how, in 1945, radio and telephone having rendered carrier pigeons obsolete, the armed forces told Whitehall's Joint Intelligence Committee (JIC) that they would no longer pay for the maintenance of military pigeon lofts. This triggered such a flurry of protest from Wing Commander William Rayner MBE, the head of the Air Ministry's pigeon section, and his like-minded colleagues, that a three-year argument within the JIC's Ad Hoc Committee ensued. Intelligence experts warned that 'pigeon research will not stand still; if we do not experiment, other powers will.'

Bee Girl didn't look very interested in the MI5 director's passion for bees, which was a shame because I'd have liked to hear more. I imagined her masked face coming towards us down the aisle in a beekeeper's outfit, fumigator in hand.

Alfred grumbled about how difficult things were getting 'for us'. He didn't specify who 'us' was. He said that 'we' had recently lost our lip-reading expert, Jessica Rees, owing to a stupid and minor error in her CV. I had read about Jessica Rees at work. She had been the foremost forensic lip-reader in over seven hundred trials, including one of the members of the South London music collective, So Solid Crew. Shane Neil, an MC with the group, was charged with drugs and firearms offences. The key evidence against Neil was Rees's transcription of CCTV footage of the moments before and after the incident. Neil, who is seen waiting with other members of the collective outside a London nightclub, removes an item from his sock. And later, according to the transcript he says: 'Do you want to buy a tablet?' After police officers move in to search him, according to Rees, he then says 'get rid of the . . . gun'. Another expert lip-reader challenged Ms Rees's transcription and told *Newsnight* she did not find the word 'gun' on the footage. The Crown Prosecution Service had recently dropped her after she was accused of misleading the court into believing she had a degree from Oxford.

'She had been to Oxford. Just hadn't completed her degree. And I mean she was absolutely fantastic. Best lip-reader we've ever had.'

The last stop was Lausanne where we got off and began our hunt for the train that would take us to the Abode of the Devil.

We found our way to a platform on the outskirts of the station. A mass of confusion flocked around two train carriages; one was roofless and contained wooden benches, the other had a plastic roof, sealed shut windows and no seats. Forty people, most with American accents, followed us, loading our luggage into the covered carriage and sitting excitedly on the open-air wooden benches. I sat on the front row, with Doug next to me, and Johnny the other side of him.

The train began to rattle its way upwards and trees multiplied into hills. A castle appeared on a hilltop, towering above us. We wove round it until it was below us and continued our ascent up the now steep mountain track into the Alps.

'It's like something out of *Deliverance* isn't it?' Doug shouted excitedly as the now thick forest darkened around us and we approached a tunnel. I was thinking about *Chitty Chitty Bang Bang* and the scene in which they descend by balloon into the land of Vulgaria.

'Who do you think will disappear?' shouted Johnny as we were engulfed by the pitch-black dampness of a tunnel.

Once out the other side, I was overwhelmed by the height we had reached and my fear of altitude sickness resurfaced. I felt very tired. I turned to look behind at my fellow passengers and my vision filled with men in shades and loud shirts and their drably dressed travelling wives. I concentrated on the woodland paths and small wooden huts. I could always run away with the fairies and camp out in one of them, I thought. The mountains were honeycombed with hiding places.

An hour later and we arrived in the valley of Les Diablerets. I stepped off the train into a wooden station building and staggered on to a road lined with wooden cabins and boutiques. I gasped at the beauty of the Alps encircling us.

Hand-painted signs indicated the way to the Glacier de Tsanfleuron and various hotels and tourist attractions. The

English headed to the bar immediately opposite the station. The Americans disappeared into the hills. There were no signs to my hotel. I followed the English to the bar, seemingly unable to muster the presence of mind to find the way to my resting place alone.

'Are you going to have a drink?' I asked Bee Girl.

'Just one. Then I'll come with you to the hotel,' she promised. She could see I was faltering.

As I made my way resignedly to the table, a middle-aged woman, who must have been in the bar when our train arrived, greeted Jeremy and Doug. There was a strange air between her and them, as if they had known and liked each other better in the past. She introduced herself warmly to Bee Girl in a southern English accent, as if welcoming a young newcomer, and ushered her affectionately towards the table she herself had just arrived at. She sat down opposite where I was slumped.

She removed her sunglasses, revealing the stern features of a bitterly determined woman and glared at me with sharp, bird-like eyes.

'Who are you?' she squawked. The high-pitched words pierced my eardrum and screeched across my tattered nerves.

'Amber.' I could barely speak I was so tired.

'And where are you from?' Her thinly drawn lips seemed to bite at the words.

'I am just a lawyer,' I stuttered and looked to Bee Girl for assistance.

'What is a lawyer doing here?' she asked accusingly.

'I am researching the legal parameters of odour detection.'

'I don't think we need a lawyer at this conference. Why are you here?'

'There are legal implications of the new forms of detection.'

Bee Girl handed Bird Woman a drink and I decided to make a move towards the hotel. I stumbled past everyone with my trolley bag and zigzagged towards a fork in the road where the hotel was sign-posted. I turned left over a pretty wooden bridge that traversed a lively stream, towards tennis courts where sweaty men and women sporting headbands waved at arriving conference attendees. The hotel was a huge and modern building with automated doors and a marble lobby containing a bar. At reception they told me they were full and that I was to be accommodated in a neighbouring hotel; it was a five-minute walk back to the fork in the road and in the other direction.

My new hotel was much smaller and old-fashioned. I opened the door to my room with a huge sense of relief to find I had a bathroom, two small bedrooms and a balcony with a view of the mountains. Perhaps this conference wouldn't be so bad after all. I set my alarm for an hour and a half later and fell asleep.

I was far from refreshed when I awoke but without a penny in the bank I needed to take advantage of the meals provided. I showered and put on a black and white buttoned cotton dress, perfume and make-up and made my way down the hill inhaling nicotine in an attempt to ignite my brain.

I arrived a little late and most places in the dining hall had been occupied. There were about a hundred people and each table sat eight. I made my way to the first table and sat down opposite a vacant seat that was quickly taken by an attractive, big-boned brunette. Her name tag read Tanya Stevens and she was one of the conference organisers. I had corresponded with her by email.

I introduced myself, having forgotten my name tag in the room.

'Oh, hi! How do you like it here?' she asked.

'It's beautiful.'

'Yeah, I'm so exhausted though. I have spent the weekend with my husband and friends in Zurich.' She rolled back her eyes. I noticed the large diamond ring on her finger. 'I'm planning an early night tonight. I want to check out the area tomorrow.'

'Yes, Les Diablerets sounds like an interesting place in the guidebook.'

'I know!' she nodded enthusiastically.

'Did you read that this place is called the Abode of the Devil and it used to be populated by elves and fairies before the devils arrived?'

'Do you think a fairy is better than a devil?' she asked, tipping her static hairdo down to the right and looking at me intently.

Out of the corner of my eye I noticed stuffed vultures and other fanged fauna peering at me through bottles of spirits on the shelves in the lobby's bar. I remembered something Cass had told me about US secret services identifying themselves with ominous mythological figures.

'I have no idea,' I said.

Officials Admit Doubts Over Chemical Plot

Guardian
Richard Norton-Taylor and Vikram Dadd
5 June 2006

Counter-terrorism officials conceded yesterday that lethal chemical devices they feared had been stored at an east London house raided on Friday may never have existed.

Confidence among officials appeared to be waning as searches at the address continued to yield no evidence of a plot for an attack with cyanide or other chemicals. A man was shot during the raid, adding to pressure on the authorities for answers about the accuracy of the intelligence that led them to send 250 officers to storm the man's family home at dawn . . .

Officials are not yet prepared to admit the intelligence was wrong . . .

A senior police source explained: 'The public may have to get used to this sort of incident, with the police having to be safe rather than sorry.'

CHAPTER SIXTEEN

New Think (Day One)

'When finally you surrender to us, it must be of your own free will.'
    George Orwell, *Nineteen Eighty-Four*

A smiling man in a caravan stairwell caught my eye. He held a guitar across his denim thighs and watched the star-studded sky, seemingly oblivious to the dozens of suited people marching through his campsite for the night. At the end of the green verge, I dropped back from the shadows accompanying me from the restaurant, allowing them to forge ahead into the concrete building. I stopped to take in the lie of the land, pausing to listen to the stream bubbling its way past me, taking in the moonlit Alpine view and wishing I was in one of the caravans parked here; with friends, and dope and booze, laughing in a political bypass tucked within the cosy confines of the mountains that had wrapped themselves around this peaceful valley.

Within the curtained walls of the purpose-built conference hall, I took a seat amongst the row of benches radiating from the raised platform at the far end as a short, red-haired woman dressed in navy blue walked up its steps. I recognised her as one of the waving tennis players. She wore her fringe clipped back on top of her shoulder length matted hair and thick foundation plastered her face. Leaning into the podium and pulling on her paisley necktie, she introduced the topic of this evening's speeches: 'Privacy versus Security'.

She confessed that she, like many scientists, hadn't been included in policy discussions about privacy up until now and so it was difficult for the scientific community to understand the concept. Cupping her hands together, she named the government official who would illuminate the concept for us and a tidily dressed blonde-bobbed policy dictator took her place behind the podium.

Her tone was calm, kind and authoritative and I relaxed into my seat sucking on my complimentary Swiss chocolate as she unravelled her solution to the privacy versus security dilemma. The only answer, she said, was for civil libertarians and security officials to stop working against each other and start working together. Smiling like an optimistic revolutionary, she looked up at the audience.

There were extreme viewpoints on each side, she explained. At one extreme, people equate a loss of anonymity with a loss of privacy. These extreme viewpoints could be compromised.

I noticed she had only described one extreme and wondered what viewpoint stood at the other end of her spectrum. Clearly 'security' meant more now than it had in the heyday of Aretha Franklin's song, but I didn't really know how.

I listened carefully as she outlined elaborate plans for making train stations as secure as airports. Complicated-looking turnstiles populated the images of small red stick men. The turnstiles appeared to be heavily embedded with sensors. She said the plans were impressive, but unrealistic. Commuters, she explained, had to be processed more quickly than current technology would allow. New thinking was needed.

They had entered into discussions with libertarians and had isolated their objections to privacy invasions. These objections needed to be integrated into public policy to gain public acceptance of security measures.

New Thinking should incorporate the principle of *consent* within the security system design. One model under consideration resembled a pyramid. Citizens could opt into high security schemes in return for avoiding routine and lengthy security checks.

'What routine and lengthy security checks?' I asked myself, before concluding that she was comparing her model with a future dystopia.

'So you could *elect*,' she explained, to be category A by providing a large amount of personal information. In return, you would be issued with identification, perhaps in the form of a micro-chip, stating that you were Category A. I wondered how often these citizens would need to update the government with personal information. On a weekly or an annual basis?

'The more people are willing to *share*,' she answered my silent query in a deliciously soft voice, 'the less hassle they will have. *Choice*,' she explained more firmly, 'will be central to the New Systems. Citizens should be able to choose between the provision of information on a voluntary basis, and physical probes.'

Physical probes?

'A good analogy,' she suggested, would be 'dental hygiene'.

'You can go to the dentist every six months,' she smiled,

revealing a perfect set of shining white teeth, 'or undergo a lotta pain when necessary.'

The audience laughed.

'It isn't a case of Privacy versus Security,' she concluded. '*Privacy* is something *security* seeks to *protect*.'

She left the podium to wild applause.

A square-bodied male security consultant with a proud and happy face strode up behind her and trotted out surveillance highlights of the year with hungry enthusiasm; a supermarket had experimented with a camera trained on the Gillette blade shelf, which was triggered by the Radio Frequency Identification tags inside the packet of razors to capture a photo of each customer who removed a Mach 3 pack. Another photo was taken at the checkout and security staff compared the two images to ensure they always have a pair.

'*Nineteen Eighty-Four*,' he explained, 'has long since passed. Emails belong to the internet service provider. You do as you wish with them.'

I could only guess at what he meant by this. It seemed to me that he was saying they weren't difficult to get hold of. As if the internet service providers were their friends.

On the topic of privacy, he said different cultures had idiosyncratic levels of tolerance. The Brits weren't bothered about privacy.

I found this comment surprising. We were one of the few countries in Europe resisting government attempts to criminalise anonymity by insisting we carry ID cards. We weren't going to take this barrage of surveillance technology on without good reason. Surely.

He said that all the Brits cared about was inconvenience. 'They don't want their journeys to work interrupted with security checks.' This struck me as a cynical indictment of British values but part of me wondered if it were true. I remembered how stressful I had found commuting with all the angry workers ramming into each other in the rush-hour. One woman had even pushed in front of me with the words 'It's a rat race, you know.' I had taken to cycling to work after one journey, on a train packed with stressed-out people worrying about getting to work in time, in the immediate aftermath of an as yet unexplained tube derailment and a *Metro* headline, which read 'Nerve gas Attack Guaranteed by Prime Minister'. The tension had been too much and I'd bought myself a bike.

The UK, he beamed, was well on the way to obtaining enough information about people to trace their movements as easily as their credit card purchases. He made an oblique reference to something he called 'an octopus card'. He must have been referring to the Oyster. British citizens, he said, would be even more willing to accede to security measures as a result of recent bombings.

A significant number of airports now had X-ray machines installed that could see through the person's clothing using millimetre wave and Terahertz 3 imaging techniques. People in the US had protested when they saw how short and fat they looked in the backscatter images. Scientists in the private sector were working on how to make the images more flattering. In England people weren't bothered by how they looked, but didn't want people of the opposite sex looking at their naked images. This sounded convincing; I knew how concerned the British were to prevent masturbation. This concern was easily accommodated, he said. People were working on devising an automated system in order to remove 'the human element'.

In Asia, there was little concern for privacy, except for Japan. In the surveys conducted, of those offered a choice between an X-ray and a pat down search, 90 per cent preferred a pat down search, which he jokingly referred to as 'state approved groping'. In the UK, the security personnel always requested a person's consent before searching them: 'You might not hear them, but they do.'

The audience laughed with the speaker and I tried my best to join in.

He talked excitedly about the potential available from combining various surveillance technologies: CCTV could be combined with call-tracing, for example. And then he said something I found disturbing. He explained that the UK was awaiting the extradition of a man suspected of involvement in the July bombings, so that we could lock him up – or hand him over to the Americans for them to do something else with him. I was surprised to hear people laughing at what sounded like a reference to state-approved torture.

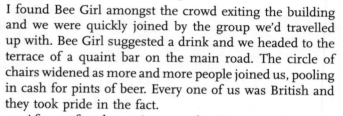

I found Bee Girl amongst the crowd exiting the building and we were quickly joined by the group we'd travelled up with. Bee Girl suggested a drink and we headed to the terrace of a quaint bar on the main road. The circle of chairs widened as more and more people joined us, pooling in cash for pints of beer. Every one of us was British and they took pride in the fact.

After a few beers it seemed quite natural to find myself talking familiarly to those around me as if I had known them all my life. Doug confided his derision of dogs and his hive-driven dreams of the future. 'We will install ourselves in the church steeple.' He outstretched his arm and looked into the distance, sparkles dancing in his eyes as he conjured up the image from his mind. 'We will work at night, using *florescence* in the bees. We will watch them from the tower, ready to raid the premises attracting them.'

Conversations ranged from the poor state of chemistry education in the UK, and the substitution of chemistry degrees with less scientifically based courses in forensic science, to the over-representation of private industry at this year's conference and fraudulent attempts by rogue private enterprises to sell governments psychic wands as crime detection tools. Drunk and exhausted back at my hotel room, the digital clock by my bed informed me that it was 3 a.m. I wanted to text Tom or Cass and tell them how weird these people were and what they were thinking. But I struggled with the wording of the text.

CHAPTER SEVENTEEN
Alien Forces (Day Six)

'Find it!' screamed The Grand High Witch. 'Trrrack it down! Rrootle it out! Follow your noses till you get it!'

The hairs on my head were standing up like the bristles of a nail-brush and a cold sweat was breaking out all over me.

'Rrootle it out, this small lump of dung!' screeched the Grand High Witch. 'Don't let it escape! If it is in here it has observed the most secret things! It must be exterminated immediately!'

Roald Dahl, *The Witches*

After five nights of drinking from 10 p.m. to 3 or 4 a.m. every night and getting up at 7 a.m. to listen to talks on bio-sensors and explosive vapours, there was only one man at the breakfast table by the time I made it there. He was an old man from the Homeland Security Department. He didn't look human and I didn't feel it. For a while we sat opposite each other eating croissants before making our way to the conference hall where I found the morning's proceedings hard to follow. The theme was trace detection. I gathered that finger pressure affected the contamination levels of particles by affecting their distribution size, or something like that. Some materials had the effect of doubling the target substance. Research was needed into the contamination caused by gloves. Some companies used the 'hacked off fingers' of corpses to handle the substances, but these weren't always readily available. A number of substances resulted in false positives for explosive and chemical weapon detection; these tended to be substances like perfume and other fragranced household products.

The 11 a.m. coffee break was a welcome respite. I grabbed a coffee, lit a cigarette and made my way to the ashtray, wincing in the sunlight as Randy Oldensaw came ambling over towards me with a smile on his face.

'Hey Amber. Good to see you here.'

'You too.'

'Hey, you know, I've got a lot more radical in my ideas since we last met.'

'Really?!' I asked. 'How so?'

'Yep.' He seemed hardly able to contain his enthusiasm. 'I think we need to concentrate our energies on stopping the terrorists *before* they become terrorists,' he said boastfully.

I laughed warmly. 'That sounds like a good idea, Randy, but how are you going to do that?'

'Kill them.'

I laughed again.

'How will you know who to kill?'

'Oh,' he pussyfooted around me excitedly, 'we have means.'

'But look what happened with the Brazilian last month in London. How can you be sure your detection methods are accurate?'

'Well yeah . . . ah . . . hmmn . . . how do I say this? Ah – a – lot has been said about that in the newspapers that isn't really true. You tell one minor insignificant lie and then you end up having to tell bigger lies to cover it up and it all gets confused.'

I didn't really follow but I could smell a rat.

'And anyway, the Brazilian guy, you know, he had been round to the houses of the terrorists . . . And he was an electrician. That's enough for me. An electrician can be very useful to terrorists.'

He nodded in affirmation of his words, seemingly confident that he'd made himself clear. Someone called out to him from the hall doorway.

'Catch you later,' he said.

I caught him again at lunch and we sat opposite each other on a table to ourselves. When I struggled to join in his monologue on the 'barbaric' and 'primitive' territories in the Middle East by mentioning the stoning to death of adulterous women, he dismissed this concern by explaining that they had a different concept of death and that this wasn't what needed fixing.

Later I got talking to a well-spoken English director of a terahertz equipment company. I asked him about the use of terahertz as a weapon and explained that Randy had spoken to me of this application.

'Oh really?' he said. 'That's interesting because I hadn't known whether to discuss that with him. No offence but, if he was willing to talk to you about it, he'll talk to me.'

That was the first sign since Bird Woman had squawked at me for being a lawyer that I'd been made to feel untrustworthy. And it seemed deliberate. After all, its use as a weapon wasn't a secret. Tom had sent me a load of articles on it. Hailed as 'the biggest breakthrough in weapons technology since the atomic bomb' it was being proposed for use in riot control. It worked by producing short invisible microwaves that vibrate water molecules in human skin, causing pain designed to equate with that of touching a light bulb. I realised that the reported effects were different to those described by Randy at the MoD conference, but no less malign. I smiled affably at him and continued on my way into the impromptu workshop on ways to improve security on the London Underground.

Inside the workshop, I was surprised by the level of enthusiasm for behavioural analysis amongst the attendees. Research was already being undertaken into the type of person who carried a rucksack on the tube. Old CCTV footage was a potential source of data for this sort of project. I thought of Garth when I heard this. He'd sold me his backpack shortly after the July bombings in London. Of Welsh upbringing, he was of Asian origins and told me he feared this precluded him from feeling safe in its presence. I'd have to tell him about his digital footprints. The American woman who'd introduced the topic of privacy on the first night was now dressed in what looked like the uniform of a Brownie leader. She stood up to make her point and addressed us excitedly, 'You said you haven't got enough dogs. Just take 'em out of the pound! They don't need to be trained! Watch how people react to their presence!' We nodded our heads. A man behind me suggested monitoring commuters' heart rates to 'detect their nervousness!'

Someone else explained the need to concentrate on unconventional behaviour, 'not Oyster card users!' Cameras could be trained to pick up on unusual behaviour like evading 'obvious camera spots'. It seemed that the security agenda involved coercing us into compliance with surveillance by making attempts to evade it inherently suspicious, like they were doing with the dogs. Their hunger for new technologies was insatiable.

There was a general consensus about the importance of concealing the weaknesses of the detection systems used from the public. But it was important for officials to remember how inaccurate most of them were, so that they were only used to trigger further alternative methods of detection. I thought about an article I had read about the recent shooting of Rigoberto Alpizar by an air marshal in Miami airport. The author suggested that the chain of events leading to the monitoring and death of this innocent man had been triggered by an algorithmic surveillance technology installed in the airport to alert authorities to suspicious bodily movements. Mr Alpizar suffered from a bi-polar disorder, a symptom of which is restless behaviour. He had waved his arms around in an agitated manner while queuing to board the plane. The article also mentioned that the vast majority of alerts triggered by trial systems of behaviour recognition technology in one London station were mistakes.

A senior transport security official, who attended the transport security workshop, addressed us in the after-dinner speech. He spoke in a sincere and anguished tone.

'I was sitting on the river today thinking over what I was going to say when an analogy came to me. We, all of us in this room, are living in a security landscape that we control. Or at least we think we think we control it. This security landscape is populated by people like us. Or is it? Alien forces have come into our landscape. These Aliens are in a human form and look like us.'

This conference was getting more bizarre by the minute. Ignoring my thumping head, I followed the herd out of the conference and ended up in the bar once again. After several rounds of drinks I ended up at a table sitting opposite Jeremy. 'I mean why not just let the government have what it wants?' he said to me. 'I think this privacy malarkey is a load of bollocks.'

'You are assuming,' I slurred, 'that the government will always be made up of good and trustworthy people.'

'Well if I thought that the people in government were bad guys then I would turn wicked.'

'Then you wouldn't want a chip in your arm.'

'I would rip it out.'

An elderly and athletic Canadian secret serviceman told me about his volunteer duties as a first aid worker at music festivals while he massaged my back at the bar. He laughed intermittently, flashing his bearded white teeth as he told me about the required shift in focus 'like your Tony Blair is advocating', away from the individual and towards community values. I sat silently on the bar stool as he rubbed my stiff shoulders, wondering if he'd form any suspicions from my growing knots of tension.

'You can come up to mine and help yourselves to my mini-bar,' Bee Girl offered after we'd been drinking under the gaze of stuffed vultures for six hours. The shutter rattled as the barman pulled it down and Doug and Jeremy bade us goodnight. The rest of us followed Bee Girl up to her room, which was plusher than mine. She served us drinks, sat down on an enormous white bed, crossed her legs and smiled. The rest of us sat on the sofa and chairs.

The chubby bloke from the MoD started telling someone in the room about a paper his colleagues had just completed on the dog's olfactory abilities.

'From a whiff to a wag?' I interjected drunkenly.

The chubby bloke flinched. 'How do you know about that?' he asked.

'Amber seems to know everything about everyone,' said Bee Girl, tipping whisky into her mouth while looking at me out of the corner of her eye.

'You see?' the bald patch on the back of the transport official's head turned to face me. 'Alien forces.'

After an awkward pause and some stilted conversation, I walked out of her hotel room into the fading night and wondered what the fuck I was doing and why I had deliberately entered their violently untrusting world. I reminded myself I was garnering information; but when I tried to piece together my rapidly fermenting morsels of memory, the conversations, meals and faces morphed into a meaningless sludge. I had lost sight of the distinction between friend and foe and desperately needed some rest. As I walked up the hill to my hotel, flute-playing goblins decorated the doors of the chalets I walked past. I watched my hands clapping at jokes of torturing terror suspects, my lips smiling at security hawks slurping soup and masticating meat and the weight of my nodding head as I listened over lunch to foreign defence officials answering my questions on the false alarm rate of the technologies we'd been hearing about.

'Ah yes, the science may be unreliable, but if we present criminals with evidence they think is infallible, then they will fold and this is a good thing.'

I crossed the stream to the road of my hotel. A bearded Alpine man ambled across the street in front of me, his eyes dancing. Young muscular wood cutters and sweet young Swiss maidens laughed in the bars. Thousands of stars wished me goodnight. Once inside my hotel room, I checked my belongings for signs of interference, decided against texting Cass out of fear of interception, and sprayed my pillows with the relaxing scent of lavender before falling into a deep sleep.

CHAPTER EIGHTEEN

Flying Lawyers (Day Eight)

At 8.30 a.m. my alarm went off and I surprised myself by getting up full of anticipation for the day ahead. I stepped on to the balcony adjacent to my bed and stretched. The sun was shining on the wooden cottage across the road and all around it was a beautiful green. I showered and dressed and ran downstairs where a buffet of cold meats, cheese, croissants, yoghurt and muesli awaited me.

The man opposite me at the breakfast table looked like someone out of a colourful dream I might have had about Las Vegas. Feeling refreshed after nine hours' sleep, I poured myself a glass of orange juice and said good morning.

His huge red face was sunk motionlessly into the

neckline of his Hawaiian shirt, its expression frozen in distaste.Having received no response, I turned to my plate. Then, hearing what sounded like a grunt, I looked up at him expectantly.

'So you're a lawyer, huh?' he said finally, turning his pock-marks to face me.

'Yes.' I smiled.

'You know what really gets me about you guys; you lawyers and judges?'

'No.' I smiled proudly on behalf of my legal brethren.

'You get away with whatever you want.' His glare reddened. 'There are limits to what I can do. I am bound by the laws of physics.'

'That doesn't sound very restrictive.' I laughed before taking a sip of my coffee. 'We have a code of ethics.'

He grunted again, heaving himself out of his chair and leaving the breakfast room. I wondered what it was that he wanted to do that he found difficult with physics alone, other than standing up. I could only think, given the hostility he had shown towards the justice system, that he wished to pass judgement.

The man next to me was more pleasant. A short Italian gentleman, he came from a long line of doctors who had specialised in X-rays of the infirm. He had inherited the family business and branched out into security. Business was good.

'So were you involved in the goings on last night?' Mandy, a woman from the Australian customs, who had adopted me as her smoking friend and had taken photos of me with her digital camera, asked me on the way into the hall. Perhaps I didn't look as refreshed as I felt.

'No,' I said. 'But I have every other night and so I decided to catch up on some sleep. Did I miss a good one?'

'Oh apparently it got pretty messy,' she said. 'The Brits

have been chiding the Americans all week for not joining in the drinking and so last night they did. Everyone got really drunk and someone got in trouble with the hotel for riding on the statue of a cow they've got in the foyer.'

I'd missed the big one. But I felt a lot stronger for a good night's sleep. I'd been in danger of losing my grip after I was identified in Bee Girl's hotel room as an alien force, and my sense of isolation and vulnerability was turned up a further notch last night when a member of a Middle Eastern security service warned the audience that terrorists came in many guises and presented a slide of a 28-year-old female lawyer. Their defence studies suggested that most terrorists weren't caught by technology, but as a result of an agent's intuition. They had carried out systematic studies into the signals agents were intuitively picking up on, but the results were confidential. I wondered how they did these studies. Presumably it was more sophisticated than the psychic experiments that the MoD had publicly confessed to. It sounded like international security was breaking beyond the parameters of the paranormal and it reminded me of the criminological interest in the ability of wild animals to foresee impending disaster. I'd sat alone on a row in the middle of the auditorium, looking up at the smiling photo of the young woman terrorist on the projection screen and started to space out. I glanced around the room. The Cheshire Cat was sitting at the front, with the man from the terahertz company; the squawking bird-like woman was behind me. My thoughts span wildly out of control as the security agent concluded his talk. Could they hear my heart beating faster and louder? Were they watching the blood fill my brain? Wide-eyed I scanned the room, conscious of my apocrine sweat glands being pumped into overdrive by my angry amygdala. My hair follicles waving frantically as the scent of my fear wafted up into the roof of my mouth, I had exited the building the moment his talk finished and put myself to bed.

'You'll come to the glacier today then, right?' Mandy nodded enthusiastically. She had asked me yesterday but I'd been scared of going up a 3000-foot precipice with her at the time. She'd been very warm and chatty with me since she'd discovered I was a smoker but she'd confessed over cheese fondue to her agency's role in informing the Indonesian authorities of her fellow citizens' plans to export heroin from Bali into Australia, instead of waiting to arrest them on Australian territory. It had caused a huge stink in Australia because the government there was meant to be opposed to the death penalty, the likely sentence for those apprehended in Indonesia. Indonesian executions are typically carried out in an isolated jungle or on a beach, where about a dozen élite paramilitary personnel shoot the prisoner in the chest as they are hooded, handcuffed to a pole. I'd found her attitude to the incident disturbing and had read that the glacier was the location of the tower-shaped rock, called the 'Quille du Diable' or 'Devil's Skittle', which was believed to be the playground of the area's evil spirits.

'Sure,' I said, no longer paranoid that she wanted to throw me off it. 'But first I have to give my poster presentation.'

I had to erect my poster during the coffee break. There were six rows of boards and posters were to be erected on both sides of each. During the session I was meant to stand by my poster and take questions from people interested in my work. Instead of one large poster, my presentation consisted of a collection of small laminated cards. I had quotes from the New South Wales Ombudsman report, a graph I'd designed on the right to privacy and a newspaper cartoon sketch of two dogs chatting about the news headline 'Drug-sniffing Dogs' Accuracy is Challenged'.

The first dog says 'Some judge has low confidence in our olfactory skills,' to which the other replies 'I know . . . I smelled it on him.'

A blonde woman was erecting her poster opposite mine. She was from a company specialising in mass spectrometry. A box of text on her poster referred to the high success rate they had had in the cases in which trace evidence had been relied upon.

'Presumably you are referring to convictions as successes?' I said to her, pointing at the box. I'd become interested in trace detection as a result of office work. In my opinion a number of people had been unfairly convicted by these libellous particles.

'Yes.' She turned smiling to face me, sticky tape in hand.

'What about the people convicted on the basis that their banknotes contained traces of cocaine?'

She glanced at my poster, which stated that I was a barrister from the Honourable Society of Gray's Inn. I don't know if she noticed the laminated cartoon of the dogs.

'Well it's a result of recent research in this area that it is now accepted that almost all banknotes carry traces of cocaine.'

'Yes, a number of prisoners appealed on the basis of similar findings. In some cases the Court of Appeal ruled that the trace evidence should not have been admitted because it lacked significance?'

'I expect so, yes.' She smiled ingratiatingly.

'Yes. Their appeals failed on the basis that in the court's opinion the trace evidence was unlikely to have influenced the jury's decision.'

'Oh.'

'Which makes one wonder why it was deemed admissible in the first place really doesn't it?'

'Well, we now advise that there must be above average

contamination levels for evidence of their existence to be significant.'

'Why is an above average level of contamination of any greater significance? We are still talking nanograms of cocaine aren't we?'

'Well it suggests that the person carrying the notes may have been dealing in cocaine.'

She seemed to be asking me to take the concept of dirty money literally. To my knowledge cocaine dealers don't mix the money they receive for the cocaine up in the cocaine before handing it over to the purchaser. Why would they? As a Colombian ritual? 'First, we must shake your money in our vat of cocaine. To show we are brothers.' Or did their very pores leak the powder on every note they touched?

We got back to securing our posters as I puzzled over what cocaine dealers looked like in this woman's mind.

A 70-year-old Ronald Reagan look-alike approached me during my poster presentation, introduced himself and told me about his work for Bush I and Bush II. His department had recently funded the opening of a brain-imaging facility, to the tune of $4 million, to study the causes of drug abuse. He couldn't understand why it was, that despite young people being informed of the dangers of drug use, they were unable to resist taking them. The new facility was going to scan young drug users' brains to find the cause of this lack of 'self-control'. It occurred to me that what he was looking for was quite the opposite. These young people, in taking drugs prohibited by the government, were refusing orders and taking part in mental mutiny. They were taking cognitive control. There was something wrong, I felt, in seeking to diagnose those seeking to exert autonomy over their neural structures as medically unsound. What would Bush II do with any findings? Operate? Medicate?

'I just wanna let you know,' Reagan said, moving his

strange rubbery face up close to mine, 'I'm on your side.'
He then backed away and stood looking at me in silence,
before giving me an approving nod, and wandering off.

Doug frogmarched his ungainly body over to congratu-
late me. Pulling me to one side and shielding us with his
arm, he discreetly informed me that the Reagan look-alike
was one of the most influential people there. He had
held senior positions in the US navy, the scientific arm
of the Office of National Drug Control Policy and the
Defense Advanced Research Projects Agency. He'd worked
on 'special ops' and 'spook stuff' throughout the Cold War,
and was now very enthusiastic about the crime-fighting
potential of the sophisticated military equipment he had
helped to create.

At the end of the poster session I ventured out into
the hotel lobby to have a cigarette. Reagan was stood at
the bar and I sat myself on a stool next to him. He started
reciting national heart attack rates to me for some reason.
Wishing to engage him in conversation, I told him that
I had heard Japan had a low heart disease rate. My dad
often told me that. Reagan agreed with me that Japan had
one of the lowest.

'They smoke a lot in Japan as well don't they?' I
said.

'Yes.'

'And they don't seem to die from it?'

'Well, we'll have to fix that,' he smiled, 'or I'm gonna
have to think up some other reason people should stop
smoking.'

He raised his glass and grinned at me, turned around
and walked out of the lobby doors as I wondered, not for
the first time that week, whether it was a mistake to engage
these people in conversation.

Mandy drove me up to the cable car station. Bee Girl was there with Cheshire Cat, one of the security consultants and a couple of young blokes with start-up surveillance companies. Bee Girl and I hadn't come into contact with each other since the drinks in her hotel room. Every time I'd seen her she'd been surrounded by admirers.

'I'm not sure I should talk to you any more if you're a double agent.' She smiled amicably.

'Who would I be working for?' I asked.

'I don't know.' Bee Girl giggled in response to my confusion. We were next in line to purchase tickets for the cable car. 'But you are the strangest lawyer I've ever met. And I have friends who are lawyers.'

Before I could formulate my response, she grabbed my arm, her face a picture of fear.

'That man has a gun,' she whispered.

I turned my head and looked at the man. He was talking amiably to the group of people around him. His right hand was in his pocket. Several white vans were parked in the car park and groups of people rushed in and out of them.

'Someone told me that they were making an action film up here,' I said. 'He is probably an actor.'

Having reassured her, I felt comfortable again in her company.

We purchased tickets and twelve of us boarded the spacious cable car. We climbed 1800 feet in fifteen minutes before the doors opened and we found ourselves on the moon-like surface of the glacier. I was standing next to the director of a start-up company specialising in surveillance equipment as we set off across the lunar landscape. He looked to be my age and his pale face was shy and friendly.

'So what is the legal relevance of the technologies being discussed at this conference?'

'Well,' I ventured, 'it's important that lawyers are aware of how these techniques work and the extent to which they produce reliable evidence. Forensic evidence can be very risky.'

'God, yeah. I hadn't really thought about it. That's how the "Maguire Seven" got wrongly convicted wasn't it? Because of the traces of nitro-glycerine?'

'That's right.'

Our boots crunched the icy snow as we slowly progressed over the terrain. A strange sort of bus drove past us. Bee Girl waved out of the window of it, laughing. We marched stubbornly along, finding ourselves alone in a crazy grey wilderness of melting snow. The others had disappeared ahead of us and the silence was ghostly.

'Some of the information you pick up on in the conversations here is disturbing don't you find?' he said.

I certainly did. Twice I'd had conversations about the recent shooting of Jean Charles de Menezes and, remembering Randy's comments at the MoD conference about terminating terrorists and the importance of reliable technologies, I'd started to imagine he'd been gunned down as a result of a false positive from one of the technologies under discussion here. Doug had said that whatever novel technology had been considered for the operation, it hadn't been used. I wondered what would have happened if it had. If our government invested in a technology that resulted in an innocent person being misidentified as a terrorist suspect on its first deployment, wouldn't they have good reason for keeping its involvement under wraps lest revealing its inadequacies would threaten our economy, our national interest? I wondered what he'd heard that disturbed him, but felt uneasy that he had decided to confide his feelings in me. My suspicions about the Menezes shooting were far-fetched and it would be foolish to share them. I didn't want to disappoint Cass by blowing my cover.

'Well, it all is isn't it?' I said.

He pointed to a crack in the ground. Peeking down I gasped at the depth of it. It seemed to go on forever to the core of the earth. He jumped up and down on it until I asked him to stop.

'That must be the Devil's Skittle.' I pointed up ahead of us as we set off again.

'Let's go,' he said, and disappeared across the white floor towards the cragged rock. It seemed we had to cross a huge expanse of emptiness in order to reach it. As we approached the precipice, a derelict shack, 'a rusty brown and deathly still chair lift, and a large bulldozer emerged from the mist.

Assuming we could go no further I stopped still and stared at the strange shape of the Devil's Skittle that towered over the precipice.

'The elves and fairies are hiding from the devil somewhere up here,' I said.

'They need our help Amber,' he said in mock sincerity.

'That's right.' I laughed.

He carried on walking up behind the bulldozer and I followed, the peaks of the world's oldest army, the Alps, becoming visible in the distance, surrounding us with their snow-coated fangs. Stairs, that had not been visible moments before, led us up to the foot of the precipice where we found ourselves on a bizarrely well-preserved terrace.

Chrome tables and chairs had been screwed into the wooden floor. The face of what I had assumed to be an abandoned building was covered in shiny metal plates and through the glass window was a fully stocked bar. Pictures of black soul singers decorated the walls. It was impossible to tell how long it had been closed but it seemed like an age. We sat down at the chrome tables on the terrace as

Cheshire Cat and the airport security man arrived. Bee Girl and the German boy came with them.

A huge eagle soared over us.

'What is that?' cried Bee Girl.

'It's an owl,' the Cat purred authoritatively. The security consultant frowned in disbelief but said nothing. Bee Girl bounced back down the stairs. I watched the bird swoop back over us and behind the towering rock. I was sure it was an eagle.

Wasps tracked by using radio transmitters

Independent
24 January 2007

Scientists have tracked the movements of wasps by attaching miniature radio transmitters to their backs. The tags, fitted to 422 wasps in Panama, transmitted to antennae placed at the entrance of 33 nests. Each time a wasp entered or left a nest, its movement was recorded, rather like people using an Oyster card on the Tube.

CHAPTER NINETEEN

Killing Birds

It was a few months before I felt ready to talk about the Swiss conference with friends. I couldn't name the speakers, because Chatham House rules required I keep their identity confidential, and I'd found it hard to account for how unnerving I'd found it without doing so. Surveillance wasn't big on the media agenda yet and anything I mentioned made me sound paranoid. But that didn't stop the muzzled memories leaping into my mind at random intervals. Although I hadn't learned anything new about smell at the conference, it seemed clear that policy advisors were looking for technologies they could use to justify hunches; or to coerce suspects into confessing; or to legitimate their use of force with something that could be labelled intelligence. Smell fitted the bill.

One afternoon, when we were enjoying a temporary lull in the number of appeals against convictions coming into the office, I told Garth a joke about bees made by Jeremy at the bar on our last night in the Abode of the Devil. Jeremy had suggested, bending forward towards Bee Girl as he did so, and leaning back again in laughter afterwards, that the sting shouldn't be taken out of their bees, as that way two birds could be killed with one stone.

'What gave you the impression he was joking?' Garth asked.

'He laughed when he said it.'

'Was it an evil cackle?'

'Well no, just a drunken laugh. It was 3 a.m. in the hotel lobby bar. I don't think it was a serious conversation.' This wasn't true. I'd found it disturbing at the time. That was why it had taken me so long to mention it to anyone. I was relieved that Garth was taking an interest in it. I thought back to the other things said on that last drunken night. Jeremy had asked me if I'd seen attack dogs in action. I lied and said I hadn't.

'You will!' He'd laughed heartily, mouth wide open and baring his teeth as the Cheshire Cat growled that I 'really should consider working for us rather than against us'. I'd said that depended very much on what work he did.

'Yes, what is it that you do?' Jeremy had smiled at the Cat, tickling him with his teasing. A strange noise rumbled through the Cat's chest as he evaded the question, smiling contentedly with himself.

'The only reason I ask,' said Garth, 'is that I am reading Jon Ronson's *Men Who Stare at Goats* and Ronson refers to a leaked document from the US Defense Department in which they discuss means of getting bees to sting their enemies.'

'You're kidding?' I said.

'No.'

Martin had advised me to look into biological warfare when I'd mentioned the bees to him. Maybe this was why.

I thought about the host of other animals I'd discovered awaiting recruitment to the security services when looking into Cass's theory that they were selected for their fear factor. At a 'top secret' location in New York State, a computer sits monitoring the movements of bluegill fish in a tank. Apparently if a fish coughs, the computer sends an alert and takes water samples. The Intelligent Aquatic Biomonitoring System (IABS), which the New York City Department of Environmental Protection has borrowed from the army, is a 'new weapon' against potential attacks on the nation's water supply. An unnamed commanding officer from the US Army Center for Environmental Health Research refused to disclose where else the army had installed the IABS. In an interview with *The Christian Science Monitor*, the unnamed commanding officer stated that the fish tanks were viewed as a deterrent by the army; 'you let them – the terrorists – know that you have this,' he said. There is no danger of the terrorists reverse-engineering the technology to foil the fish; 'the beauty of it is that you have to understand what the bluegill can and cannot pick up, and the key feature there is that even I don't know that.' But surely fish weren't harmful? Unless attacking other fish?

Russian airline Aeroflot had cross-bred jackals with dogs and described their progeny as a 'psychological weapon'. Aeroflot believed that the dogs, which are expensive to breed, will pay their own way as a large amount of interest has apparently been expressed in them by other airlines. Aeroflot was planning to market the expensive hybrids worldwide. Their motivation seemed to be financial.

Then there were the pigs. According to Yekutiel Ben-Yaakov, director of Jewish Legion, a group responsible for providing specially trained canines to settlements to serve as 'defensive guard dogs trained to subdue, rather than kill' intruders, pigs were being trained up to sniff out weapons and explosives from great distances, and then to amble towards the alleged terrorists, thereby identifying them to human guards. The director explained that according to Islamic belief, any terrorist who touches a pig or even pigskin 'would be denied the reward of 70 virgins in heaven'. This perceived danger of the pig was sold as an additional advantage to their use, but it hardly amounted to biological warfare.

Parasitic wasps had also reported ready for duty. A tank, or portable device, as it was referred to, had been especially designed for them. It was the size of a cup and called a 'wasp hound'. A small camera communicated the wasps' intelligence reports to a computer, which translated them into flashing lights. Wasps, I imagined, could be dangerous.

But bees, it seemed to me from the intelligence I had gathered, were the future of law enforcement. It had only recently dawned on me, as a result of my reading of Lt Col Richardson's *British War Dogs*, that the growing army of police dogs was in fact a peacetime reserve for the military. Richardson, who had dedicated his life to persuading the military, law enforcement or indeed anyone who'd listen to make greater use of his personally trained canines, failed to persuade the army to train dogs in advance of the First World War despite his reported discovery of 'German agents buying up large quantities of our good Airedales, sheep dogs and collies, for military and police purposes'. When the war came, and the German army was discovered to have thousands of war dogs at its disposal, it was too late for Richardson to supply enough trained dogs to meet the British army's needs. He later recommended that a peacetime reserve of dogs be maintained by the police force, and their services could then be called upon in the event of further hostilities. Could this be what the bees were for? I was desperate to find out but I'd have to wait.

'Thanks Garth. Perhaps you could let me have a look at it when you've finished. I've been meaning to look into bees some more anyway. I've got to do this counter-terror thing first, though.'

Garth laughed. 'I still can't quite believe you are down on the brochure as a security expert. I mean you, a counter-terror expert? Have you worked out what you are going to say yet?'

'No.'

'Maybe you could talk about the bees.'

'Ha ha.'

I'd been so surprised at the plans that were being made for our future at the Swiss conference that I'd wanted to find out more. I'd also wanted to check my security credentials were still in order after it. I'd not heard back from Bee Girl or Doug, despite sending them a number of friendly emails on my return. So I'd applied to attend a counter-terror conference. Having let it slip in my application for a place that I'd given a presentation to the Ministry of Defence and to international security representatives in the Abode of the Devil, I was invited to sit on their expert panel. I could hardly refuse.

Colourful flyers for the counter-terror conference event had been posted through my door and I was encouraged to advertise my upcoming performance to colleagues and business associates. Unfortunately the topic for discussion was business and counter-terrorism and all I could glean from the relationship between the two was that counter-terrorism was good business. I couldn't even fathom what was meant by terrorism or by a secure world. This conference, like many others I'd considered attending, viewed natural disasters such as earthquakes and hurricanes in the same light as terrorist attacks and criminal predators. They threw them all in together and claimed to be able to save us from them all. In the newspapers and on the radio, politicians prattled about the right to life, which somewhere along the line they seemed to have equated with a right to physical immortality. And I didn't understand the way politicians had started referring to 'the environment and security' as the reasons behind their every move. It struck me as a potentially menacing combination.

The title of the conference was 'Building a Secure World' and the image for my session portrayed the gleaming green wolf-like eyes of a man inside an ominously red astronaut's mask. An appropriate outfit, it struck me, for the risk averse, but far from the comforting image one might expect for a safe planet Earth. It looked like the world these technophiles were building wouldn't be as safe as it was secure. In fact the whole thing struck me as slightly absurd. The list of attendees comprised high-ranking politicians, government security personnel including the director of defence security at the Ministry of Defence, senior police officers and PR and security representatives from Wal-Mart Stores Inc., Asda Stores Ltd, BAE Systems, Volvo Technology, Lockheed Martin, Vodaphone and British Airways, amongst others, along with a whole booklet of security companies that would be present to advertise their wares in the exhibition hall.

Passengers warned they will face tougher airport security checks

Independent
Johnathan Brown and Nigel Morris
11 August 2006

Air passengers were warned that they face far more stringent security checks in the future as experts accused the Government of delaying the introduction of new explosive-detecting technology.

. . . A range of new products designed to combat the change in terror tactics had been rushed to market in the wake of the attacks in America on 11 September 2001.

Smiths Detection Systems has provided 6,500 of its 'trace detectors' worldwide – half of them in the United States – which can 'sniff out' remnants of explosives and drugs.

CHAPTER TWENTY

Eden Olympia

Security is a false god; begin making sacrifices to it and you are lost.
   Paul Bowles, *Notes Mailed at Nagercoil*

Most of us on the train pulling into Kensington Olympia were probably conference attendees, but the brazen indiscretion of the man opposite me was still a shock to the system.

'The government's still refusing to sign the contract. It's time to play hardball. We need to get the WTO involved, get them to put the pressure on,' he said into his mobile phone.

The train doors opened and I tried to follow him to find out what government he was referring to, on behalf of which organisation he was talking and what kind of pressure the World Trade Organisation could exert, but I lost him in the crowd exiting the station.

The security on the way in to the exhibition hall was predictably tight; the event was sponsored by Group 4 Securicor, Smiths Detection, Thales, Security Management Today, Mannix Security, BMT Centre for Homeland Security, Oracle Corporation and L3 Communications.

The ground floor of the exhibition hall was a hive of activity. The Department for Trade and Industry handed out brochures outlining the assistance it could provide to security companies trying to get a head-start in the competitive industry of counter-terrorism and a woman in black leather hotpants wandered up and down the hall with Love Heart sweets, inviting visitors to see the security solutions she could service them with. Numerous companies advertised their wares with the slogan 'The Surveillance of Tomorrow Today' and others asked 'If the nation looks to you for its security, who do you look to?'

Patrick Mercer, the then shadow home secretary, addressing an audience filled to the brim with security company executives, remarked on the unprecedented move by Eliza Manningham-Buller, the then head of MI5 and reputed bee-keeper, to inform the public of the gravity of the terrorist threat faced. He thought that was a good move. In fact he thought that the government should do more. The public, he said, should be kept permanently aware of the level of threat faced by the UK of terrorist attacks. The threat level should be regularly communicated in cinemas and on the radio and television, as it was in the US.

'Who knows,' he concluded, 'how big the next threat will be? There is no way we can tell but the number of casualties will be in the thousands.'

Private industry should be given financial incentives to incorporate security systems within their architecture.

The director of Transport Security and Contingencies (TranSec) assured the audience that the government was continuing to invest in counter-terror technologies and was conducting screening trials of new technologies at stations to see how well they worked and, importantly, the extent to which the travelling public was willing to accept them. The results so far were encouraging. The vast majority of the travelling public in the UK were happy to comply with the security arrangements. In fact, she'd been surprised by how few refused to undergo security screening. The results suggested that the public accepted the necessity of security measures, which democratically legitimised government moves to install them on a more permanent basis. I thought back to the encounter I'd had with the mobile police scanning equipment in Seven Sisters station the week before. A black arch had come into view as the electronic stairs escalated my exit. People in front of me walked through without batting an eyelid. I stopped still at the top of the escalators before going through it. Police tape extended from each side of the escalator to the arch. Green lights flashed on either side of it. It looked like a metal detector, but I couldn't be sure. Three people in London Underground uniforms, the new blue ones with capped hats, stood to my right.

'Excuse me,' I'd called out to the nearest person. Her eyes met mine uncertainly.

'Do I have to walk through this?' I asked, as the set-up appeared to offer me no alternative.

'Yes,' she said.

'Says who?' I asked.

'All those police officers.' She pointed out four Metropolitan police officers, standing with their arms crossed through the arch's window.

I walked through the arch and up to one of them.

'May I ask under what powers you are obliging commuters to walk through your scanner?'

He said he didn't know and told me to ask the sergeant who was standing next to him. The sergeant told me that they were Metropolitan police officers and that they were conducting this operation on behalf of British Transport. They were allowed to do it, he said, because the station was 'within British Transport jurisdiction' and it was 'private property'. This didn't make any legal sense to me and it still doesn't. I explained that I was a lawyer and well aware that the police did not have a power to stop or detain me in order to find grounds for a search. I understood that the official police line was that my consent was meant to be obtained before scanning me. The British Transport police press office had told me so when explaining the difference between it and a dog sniff.

'Well you didn't have to walk through it,' he said.

I turned round and looked again at the police tape.

'How else could I have exited the station?'

'You could have walked round the side of it.'

'What, under the police tape?'

'I wouldn't have said anything if you had. Or you could have said you didn't want to walk through it. Obviously,' he added, 'that would have given me reasonable grounds to search you.'

'What for? Failing to comply with a voluntary procedure?' I knew that behaving in a manner that is uncooperative, but not inconsistent with legal rights, cannot in itself provide the police with reasonable grounds to search.

'Yeah. Shows you've got something to hide.'

'No it doesn't. It shows that I expect the police to act within their powers.'

'What about security?' he asked.

'I think that's best achieved if we all act in compliance with the law, including police officers, don't you?' I asked.

I thought about quoting John Locke at him; 'wherever law ends, tyranny begins', but I could tell from his face that he already disliked me on account of my profession.

After he'd shrugged my question off dismissively I left the station. If only we had an equivalent of the US Flex Your Rights Foundation, a public education group teaching people to understand, appreciate and assert their constitutional rights during police encounters. I might have been more prepared for the situation.

There certainly hadn't been any suggestion that I had a choice about compliance. I had no idea if this had formed part of the TranSec director's pilot studies but I'd received reports from a number of friends and acquaintances about mobile police patrols they had come across and they all sounded similar to my experience. And at the Swiss conference a representative from one of the companies piloting the see-through-your-clothes people scanners in Heathrow Airport told me he was uncomfortable with the official line that people's consent was being obtained. 'I'm not sure it's right to call it consent,' he'd said. If it wasn't made very clear to travellers that they were under no obligation to comply with new security measures and that non-compliance would not arouse police suspicions, then I couldn't see how their compliance amounted to public approval of them. And even if it did, this wasn't the traditional method of law-making; we have a legislative process. I recalled a conversation I'd had with a transport official late one night at the Abode of the Devil. I'd ridiculed him for his reliance on focus groups, referring him to Chris Morris's spoof documentary for Channel 4 in which he got together a focus group and asked them how many of them would support the insertion of a pea-sized object within a paedophile's anus that would expand to the size of a television set when he came within a certain distance of a child. The vast majority said that they would. He'd laughed and said he'd mention that at the next departmental meeting.

The director of TranSec confirmed that the threat to the

UK was real and serious and that international terrorism was here to stay. It sounded like the War on Terror would be a permanent state of affairs and the audience seemed pleased. The UK government focus, she stated, had to be on research and development of new security technologies.

A gentleman at the back put up his hand and addressed her.

'What about canine assets?'

The director looked confused.

'Canine assets are more effective and reliable than electronic devices. Why aren't you investing in them?'

The director still looked confused. Someone else in the audience explained that he was talking about dogs. She smiled kindly at him, and said that the Department for Transport was well disposed to canine assets and agreed in principle with using them.

I cornered the gentleman on the way out of the talk. He was fuming. He ran a private dog security company and he was pissed off with the Department for Transport's 'preference for technological gimmicks' when canines were more accurate and more 'culturally significant'. He attributed it to political lobbying by security firms and he thought it endangered our security. To my surprise, I found myself briefly agreeing with him. In the months that had passed since my visit to the Abode of the Devil, I'd noted ion-track scanners, a trace detector of molecules of interest, vying for the role of a policeman's best friend. It was more expensive, costing up to £40,000 a unit and £1 a swab, and appeared to be equally unreliable, if not more so, in its ability to correctly identify illegal drugs, but it worked well as a police tool for increasing their powers without recourse to the law. It was wheeled round licensed premises, erected in club doorways and taken round festivals like a circus trick. As with many of the police toys that haven't been authorised by legislation and exist outside of the regulatory framework for surveillance, it is described as a 'voluntary' procedure. Members of the public are asked if they are happy to be swiped with a swab. A refusal is frequently treated by the police as grounds for suspicion justifying a physical search, by-passing the law on stop and search and reversing the burden of proof in a blinding twist of Kafkaesque logic. It is well advertised on the American conglomerate General Electric website:

> Invisible traces of narcotics or explosives left behind on keys, drivers' licenses, door handles, jacket pockets and other high-touch areas are often the key to obtaining a search warrant that enables officers to stop a crime before it happens.

General Electric is only one of a number of suppliers of this machinery to law enforcement. As well as investing in the machines themselves, UK police forces are persuading nightclubs to buy them by telling them it will help them to keep their licences. Many club owners are angry about this because the machinery is very expensive and its use results in long queues. At first I'd found it strange that clubbers queued up to be tested, but I'd learned that the club would be surrounded by dozens of spectator police officers on the look out for 'avoidance behaviour'.

But the most bizarre thing about the use of the technology is that its false alarm rate is hailed as an advantage on the BBC News website:

> The efficiency of a high-tech drug-testing machine unveiled in Britain was amply proven when the government minister showing it off tested positive for cannabis ... the politicians were keen to stress that such was the power of the device, positive results could easily come from so-called

'cross contamination', for example by touching cash or a door handle previously handled by a drugs user . . .

Police explained later that while a positive test could not be used as evidence in court, it could help police to target people to search or question.

Some police officers had even complained to me about the government targets that they felt obliged them to resort to such tactics. Unlike in Spain where the Mallorcan police held a sit-in protest when the government asked them to increase the number of arrests, the British police have no right to strike. I tried asking a couple of the ion-track providers, who had stalls here, about the false positive rate and they had all responded cagily. At first I couldn't understand why. I'd explained that this information should be public knowledge because the police were using its results as grounds for searching people. Then one of the company reps took me aside to explain that there was a large amount of corporate espionage going on, and no one was giving that sort of information away.

Having said all that, my heart didn't go out to the gentleman from the private dog company. Judging by the number of private dog detection companies represented at this event, I wasn't convinced that the canine business was doing that badly out of our heightened security measures. And a number of police dog handler friends that I had made recently described private dog operators as 'cowboys' in it for a quick buck and without any ethical qualms about sending badly trained dogs out into the streets.

I wandered round the stalls, filling my bag with security literature and eavesdropping on conversations. CCTV user groups discussed 'extending the rural reach' and the technology behind the UK's national ID card scheme was trumpeted as 'so advanced that no one knows if it will work'. And the gas mask supply was 'all there. We just need something to trigger the demand.' A camp and deeply suntanned gentleman in a suit demonstrated his svelte figure to me in the millimetre wave people scanner. He'd offered me a spin when I approached.

'No,' he'd looked me up and down and smiled when I turned down the offer, 'ladies don't normally want to. I'll show you how it works.' That seemed to be how he spent his days, spinning his torso on a computer monitor for security consultants to admire.

Licensed premises were giving the police direct access to their CCTV systems and as a result of 'mutually beneficial partnerships' between the security services and private industry, private companies were gaining access to large amounts of personal information. I could only guess at how useful this might be for targeted advertising.

During the discussion group on global cooperation, there was a great deal of concern about the number of years the erection of 'Total Systems' would take and a number of complaints from security consultants who had found themselves wrongly blacklisted in the US and the huge bureaucratic nightmare that had beset their struggle to reinstate their security status. I couldn't quite work out what these 'Total Systems' entailed. There was talk of 'future-proof solutions', sensors and 'airborne systems', lasers and 'unmanned vehicle solutions', electro-optics, infra-red and wheelbarrow revolutions.

In the midst of this bizarre barrage of weaponised bureaucracy I had my say on the panel. I simply noted that there had been a lot of talk about the need to balance Western values against heightened security, to ensure that terrorism didn't succeed in destroying our way of life, but not much talk about what these values were.

The only Western value I had heard mentioned was prosperous business, and the requirement for heightened security was good business for most of them here. No one seemed particularly surprised by what I had to say, though a couple of gentlemen came and thanked me quietly afterwards.

Funny, I thought, as I boarded the train home, that in the light of all we'd heard today about the likelihood of a terrorist attack on the tube, so many security analysts were happy to get on it. As the train progressed deeper into the city, it filled with commuters returning home from their office jobs and the suited men clutching their Secure World bags of merchandise brochures blended into the masses before quietly disappearing.

Guantanamo cell is better than freedom, says inmate fighting against release

The Times
Sean O'Neil
31 July 2007

An inmate of Guantanamo Bay who spends 22 hours each day in an isolation cell is fighting for the right to stay in the notorious internment camp.

Ahmed Belbacha fears that he will be tortured or killed if the United States goes ahead with plans to return him to his native Algeria.

CHAPTER TWENTY ONE

Bees are the Future

The conference dealt with, I delved into bee-world. Professor Jerry Bromenshenk, chairman of the Bee Alert Technology company, was reported to have trained bees to track landmines and locate improvised explosive devices. In an interview featured in the *Govexec Daily Briefing*, Bromenshenk explained:

> We could train them just as easily to find meth-amphetamine labs, dead bodies, or any number of other uses. Maybe even oil. Bees are the future. The defense appropriations budget for fiscal year 2007 is roughly $468 billion, and we only need $5 million to develop a species of superbee into a

complete new defensive weapons system for this country. If we could get funding from the Army, it would be great for Homeland Security, and great for the Montana economy too. The bees are not only attracted to the landmines, but they also buzz around the guys who handled the explosives, and try to land on their hands and sting them. When you're looking for insurgents, the psychological influence of that could be enormous.

I looked up bee-keeping and came across hundreds of bee-keepers' home pages. At first glance, bee-keepers, as Bee Girl had intimated to me on the plane, were a pretty weird bunch. Nearly every web page consisted of photos of their keepers standing smiling next to huge colonies of bees. It struck me that they must be similar in character to the Home Office scientist who gave CS spray the all-clear after receiving blisters to her face on being sprayed, as I read one bee-keeper's advice on whether bee-keeping was for me:

> You need to know, one way or another honeybees are going to sting you – no matter how well you dress, they always find their way inside your suit or mask . . . Bees let you know they are not happy and are willing to defend their home at all cost. They'll smack into you, sting you, buzz you insatiably and do everything possible to annoy you and freak you out.

It wasn't for me and I was surprised when I found out how common bee hives are, even in London. The number of CCTV cameras in existence pales into insignificance in comparison. The country is filled with bee-keeping enthusiasts and modern man-made hives make it 'as easy for the beekeeper to monitor and manipulate his livestock as if pulling and reading a file from a drawer'.

Holley Bishop's *Robbing the Bees* was a treasure trove of facts. Napoleon embroidered the bee into his coat of arms as an emblem of power, immortality and resurrection, and bees were described as 'an ancient, powerful and noble fighting force'. The American military came face to face with them in Vietnam. Bishop includes an extract from an interview with a Vietnamese guerrilla fighter about his use of bees in jungle warfare:

> We studied the habits of these bees very carefully, and trained them. They always have four sentries on duty, and if these are disturbed, or offended they call out the whole hive to attack whatever disturbs them. So we set up some of these hives in the trees alongside the road leading from the post to our village. We covered them over with sticky paper from which strings led to a bamboo trap we set in the road. The next time an enemy patrol came, they disturbed the trap and the paper was torn from the hive. The bees attacked immediately: the troops ran like mad buffalo and started falling into our spiked punji traps. They left carrying and dragging their wounded.

'Both bees and gods,' Bishop writes, 'seem to attract types devoted to quiet observation, contemplative thought, and a tireless search for answers.' The same, I realised, might be said for surveillance officials and detectives. At first it had struck me as strange to find so many animal keepers, entomologists, bee-keepers and zoologists in the world of surveillance. Then it dawned on me that there was a close affinity between David Attenborough's special effects and new surveillance techniques. They permit intimate observation of life, which one can follow from a distance, speculate on and interpret; and perhaps pass judgement upon in search of meaningful behaviour. Perhaps those scientists, who spent their youths studying and manipulating basic organisms before being recruited to assist the military and law enforcement in their wars on terrorists, drug users, criminals, and antisocial humans, retained the same perspective on life as that which they had developed during their university studies. In *His Last Bow*, Dr Watson tells Sherlock Holmes that 'we heard of you as living the life of a hermit among your bees and your books in a small farm upon the South Downs'. In reply Holmes shows off a book he has authored, *Practical Handbook of Bee Culture, with Some Observations upon the Segregation of the Queen*, saying, 'Behold the fruit of pensive nights and laborious days when I watched the little working gangs as once I watched the criminal world of London.'

I looked up 'bees' and 'weapons' and found myself on the website of the Sunshine Project. The Sunshine Project is an international non-profit organisation working 'against the hostile use of biotechnology in the post-Cold War era'. Hoping to strengthen the global consensus against biological warfare and to ensure that international treaties effectively prevent development and use of biological weapons, it makes carefully targeted and cleverly drafted Freedom of Information (FoI) requests and publishes the results. The documents received are sometimes heavily censored, and often disclosed only to be quickly followed by government attempts to retrieve and re-bury them; or, when that fails, to publicly dismiss their relevance.

The Sunshine Project had, by this process, uncovered a 1994 US Air Force Proposal entitled *Harassing, Annoying and 'Bad Guy' Identifying Chemicals*, suggesting the development of four classes of chemical weapon. The first is chemicals that 'attract annoying creatures' to the target's position and 'make the creatures aggressive and annoying'. The candidate creatures listed include 'stinging and biting bugs, rodents, and larger animals'. The document failed to provide any examples of the larger animals that might be useful. Lions and tigers and bears leapt to my increasingly anxious mind. I'd have thought moths and mosquitoes would be the easiest to attract by means of pheromones but the proposal appeared to favour bees; 'a "sting me/attack me" chemical that causes bees to attack would be especially effective'. The second category of chemicals proposed would make 'lasting but non-lethal markings on the personnel'. A number of variations under this category were discussed, the most 'subtle' of which was 'some lasting chemical marker that was not obvious to humans, but would be obvious to trained dogs or special detecting equipment . . . Marked individuals would not know they were marked, and would not know how the dogs/equipment identified them.' I thought about the accounts I'd received, from innocent people who had been stopped by police dogs. Were they 'marked individuals'?

I found the third category of chemical weapon equally unnerving: 'Chemicals that affect human behaviour'. It included chemicals that would adversely affect discipline and morale by inducing homosexual behaviour in heterosexual men, making the target 'very sensitive to sunlight' or giving the target 'severe and lasting halitosis'.

The Sunshine Project recounts how, on their publication of this document, the Department of Defense and the Joint Non Lethal Weapon Directorate (JNLWD) sought to distance themselves from it by claiming the proposal had not been taken seriously or considered for further development. This, the Sunshine Project maintains, is untrue. 'In fact, it was recent Pentagon consideration, in 2000 and

2001, that brought this document to the Sunshine Project's attention and resulted in our FoI request.'

As recently as 2001, the JNLWD commissioned a study of 'non-lethal' weapons by the National Academies of Science (NAS). JNLWD provided information on proposed weapons systems for assessment by a NAS scientific panel. Among the proposals that JNLWD submitted to the NAS for consideration by the nation's pre-eminent scientific advisory organisation was *Harassing, Annoying, and 'Bad Guy' Identifying Chemicals*.

I immediately contacted a company working on sniffer bees in the UK and explained that I was researching forensic odour detection and would appreciate meeting up with them. A female entomologist working for the company invited me to their laboratory.

I didn't inform her when arranging our meeting that what I really wanted to know about was the use of bees as biological weapons, but it slipped out on arrival. She said she had no idea how to train a bee to target a person and get it to sting them. In fact, she seemed shocked that anyone would even have considered it. No doubt the military didn't always inform scientists of what they were planning on doing with their research findings.

Her colleague chipped in that if that's what they'd been after, they'd be using wasps. 'Wasps can keep on stinging. When the bee stings, the stinger, poison sac and several others parts of the bee's anatomy are torn from the bee's body. Sure, the sting is painful because it's got barbs on it, and if it's not immediately removed, the reflex action of the muscles attached to the sting drives it deeper into the skin where it discharges poison from its sac. But once a bee stings,' he shrugged, 'it's dead.' Kamikaze chemical weapon pilots. No wonder the military is so interested in zoology.

I tried to think up other questions to ask them as they

seemed to think my first one was off the rocker. We were sitting at the table with a box of bees between us.

'How are they trained?' I realised was a better way to start the conversation.

'It's just Pavlovian conditioning,' she said, shifting the box of honey-bees across the table towards me. 'We blow the odour we want them to indicate on into the box, and give them sugar simultaneously. As a result they learn to associate that odour with sugar, and extend their tongues when they are exposed to it. You can do this with any smell – even compounds that would never be found in nature. It takes just a few minutes to train a bee. It's really quick.

'We developed this prototype.' She picked up the box. 'It's portable. Also it enables us to isolate the bee from any other stimuli. This is important as you'll know from your research on dogs. With dogs a lot of it is about the relationship with the handler. With the bees in this box, it's only about the odour.'

'Why three bees?' I asked. There were three in the box, like the prophetic pre-cogs in Spielberg's film *Minority Report*.

'Three is better than two!' she laughed.

'How do you catch them?'

'We catch them with a dust buster that we have modified for the purpose – it is very easy to catch them.'

The box was rigged up to a computer monitor that registered the bees' head movements.

'The length of the tongue is measured on the screen.' She pointed at the monitor.

'Do they behave any differently after their time here?' I asked, picturing the bees returning with expectations of respect from their colleagues on account of their security secondment.

'No, because they don't do anything here that they don't do in nature. We are just using their natural abilities. They go to what they recognise as a food source and get sugar from it.'

'Have you experimented with taking a sample of the substance that the bees have been trained on, say, cocaine, and placing it near the hive after their release, to see if they are attracted to it? Would they encourage the rest of the colony to collect it for an instantaneous sugar rush?'

'No, we haven't tried that but I don't really think that would happen. I think if you wanted a bee to detect a target substance in a field, you would need to train it as a free-flying detector. A lot of people use the bee as a model to study the brain and extrapolate lessons for the mammal. Because it's a simpler organism. Sure, the receptors are on the antennae rather than in the nose, but they work in the same way. It's the same basics of learning and memory that operate in other species. They don't extrapolate lessons learned in one context to another. If they come across an explosive and get no reward, they will forget it as a food source. It's a matter of experience.'

'I don't think our trained bees would pass their lessons on to the next generation.' She laughed, in response to my question about Pavlov's belief that conditioned organisms could pass their skills on through successive generations. 'We are not exposing the colony to something. We are not training the colony. We are training individual bees. We are not getting at the life of the colony, which is what might trigger evolutionary changes. With one bee, there is no possibility of this happening.'

'If,' her male colleague added, 'we were to put the smell of the target substance in the food, then bees returning to the hive would teach the other bees to search for the target substance by spreading its smell in the process of distributing the food amongst the hive. But in the lab we keep the target substance and the sugar separate from each other, so that when they go back to the hive – it's just sugar that they are carrying.'

'The US military seems to be interested in free-flying bees instead of portable devices, though, doesn't it?'

'Yes, that's the idea with landmine detection. They've awarded funding to Jerry Bromenshank to do that. They do put the sugar in the target substance.'

'And would you be OK with doing that?'

'Yes – we could do that. We are talking with the Home Office about using free-flying bees. We have never worked with free-flying bees, but we could do it. We can use any honey-bees anywhere.'

'Do you think there are any ethical issues you should be concerned about?'

'We don't kill them. We take good care of the honey-bees. The fact is that at the moment we take them for granted, they are in nature, and we don't care about them and look what happens – bees across the world are dying as a result of the chemical pesticides we use. But if we can find applications for the bees then we will probably have more of them around. We would take better of care of them. So actually I think our work is better for the bees.'

This struck me as an interesting idea: saving bees from extinction by making them valuable to the security services.

'I suppose what I meant was whether you have any concerns about the militarisation of biology.'

'I don't think genetically modified bees would achieve anything,' she said, assuming that militarisation involved more than training. 'I think it is interesting to study their genes to learn how they smell and so on. The bee's brain has been a subject of intense study for many years. But I don't think you would create a superdetector bee. I don't think there is any point. They can do the job anyway.'

'Do they all behave perfectly in the hive? Does it work

with mechanical efficiency? I've heard each hive has police bees.'

'You do have policing in the colony – but that is more about motivating lazy bees – and making sure that everybody is working.'

'So could an individual bee rebel?'

'Well a colony might behave intelligently but an individual bee acts more like, well – it doesn't think about how it wants to behave – it's just a reflex action.'

'Haven't there been some studies recently suggesting that bees do make use of cognition, that they are more intelligent than once thought?'

'I don't like to say that bees are not intelligent – it depends on what you mean by intelligence – it is a sentient system – the brain hasn't got as many neurons – you can't say that it is not intelligent – but, I mean,' she said, evidently worried that I might be an animal rights campaigner in disguise, 'people often ask me if the bees are OK, if they are stressed and my answer to that is that if they were stressed they wouldn't be able to learn.'

'Really? But didn't Pavlov say that it is easier for an organism to learn new behaviours when it is stressed?'

'Well yes, I guess that is true, because when you are in a life-threatening situation you are more alert. So I guess it is probably true but we don't deliberately stress them. I think that their life in the laboratory is very similar to their life in the colony. They work all the time in the colony. In fact they live longer here – normally they don't live in the winter but here we have a hive to keep them going with artificial light. And they have no predators in the laboratory.'

It sounded like a secure world.

'What about destroying their sense of achievement and sense of mission?'

'As long as they get their sugar they are happy. I don't think they get frustrated!'

'Why do bee-keepers use smoke when they want to inspect the hive drawers?' I'd come across numerous explanations for the effectiveness of the bee-keeper's smoke machine in placating the bees to obtain their tacit consent to inspection. The most popular, according to Holley Bishop in *Robbing the Bees*, is that 'it alarms the inhabitants, which suspect a forest fire and then hide in the combs of the hive, bingeing on honey until they become passive with sugar and fear'.

'The smoke is to mess up their sense of smell,' she explained, 'to cloud their ability to trigger an alarm signal amongst the colony. Or they think it is the smell of fire and so their natural response is to feed because they may have to flee. And when they are full they are not so aggressive.'

'So there was a big buzz about your technology a while back when it was reported that the bees would be deployed within a year. What happened?'

'Well, we can be ready in a year if we have the money. We had this big contract with the US military. First we went there to demonstrate and then they wanted to do it themselves and so we supplied the boxes. I think that they did a lot of experiments and spent a lot of money on that but they didn't want to collaborate. They don't believe they could use the technology in Iraq or Afghanistan because it is too difficult to deploy. But they are still researching this area.'

In the library I read that the latest research on the life of the honey-bee revealed how the Queen Bee exerts her power over the hive by brainwashing its members with the Queen Mandibular pheromone. This ensures that the hive remains a 'haven of order, peace and harmony'. The Queen Mandibular pheromone consists of a complex cocktail of chemicals that prevent the drones from attacking her and regulate the behaviour of the entire hive. This, the authors of the research paper concluded, 'ensures that this important survival tool can benefit workers and contribute ultimately to the survival of the colony as a whole'. Aromatic chemicals had been shown to have a wide number of applications in bee life, including commands to flee and calls for reinforcement in stinging attack. The alarm pheromone is a mix of many compounds. It smells like bananas and is released with the sting. Benzyl acetate causes flight behaviour, whereas l-butanol, l-octanol and hexyl acetate cause the recruitment of more bees from within the hive. The pheromones stay on clothing, so it is important to wash your clothes if you get stung. Research into the use of odour signals by bees had caused some scientists to doubt the importance of Karl Von Frisch's pioneering work on decoding the dance of the honey-bee as a form of intra-bee communication. It was possible that bees in fact communicated the location of a new hive or of a new food source by laying aerial scent trails and that their every move was regulated by olfactory signals. This, I thought, would make the job of police free-flying bee handlers a lot easier as there would be no need for officers to learn the waggle dance moves.

US Sets Threat Level at Highest Point Ever

Independent
Andrew Bunconbe
11 August 2006

America's security alert on aircraft from Britain was raised to its highest level for the first time yesterday – a month before the fifth anniversary of 9/11. As America woke up to the news that an apparent plot to destroy aircraft flying from London to the US had been foiled, officials from the president downwards reassured the public . . .

Raising the threat level for flights from Britain to 'red' requires airlines to give the government in advance the names of passengers headed to the US . . .

Of course, most in the West have little option but to take at face value the words of politicians with access to plot details. Yet in recent years there have been many claimed threats that proved unfounded . . .

Earlier this year, the media watchdog Fairness and Accuracy in Reporting highlighted White House claims about the 'disruption' of a plot to attack buildings in Los Angeles that intelligence officials later said had never got beyond speculative talk.

CHAPTER TWENTY TWO

For Whom the Bell Tolls

'So how do they train these bees then?' Garth asked, tapping his fingers on his tidy desk. It had taken him long enough to ask. I'd been waiting all day as he soldiered away on his files, barely acknowledging my existence. It was almost time to leave and he'd shut his computer down and cleared away his papers.

'Pavlovian conditioning. Just get the organism to associate the target smell with a reward,' I replied, putting down my work. 'You familiar with it?'

'Yes.' Garth nodded. 'Big concern in the 1930s when they started applying it to humans.'

I glanced up at the clock. Three minutes to go before his departure time.

'Really?'

'Yup. Bloke called John Watson conducted some ethically dubious experiments on children. Taught some little kid to fear rats and then not to fear them again. Albert, that was his name.'

'Whose name?'

'The child's. B. F. Skinner is the most famous Behaviourist though. Professor of Psychology at Harvard between the 1950s and '70s. Said Pavlov's methods could be applied to humans for the greater good.'

'Behaviourist?' The term rang a bell and so did the name B. F. Skinner. He'd designed the pigeon-guided missile Martin told me about at our meeting at the Dog and Duck. And Gilbert Ryle, who had coined the term 'ghost in the machine' to describe what he viewed as the fiction of the human soul, had 'Behaviourist leanings'.

'Their school of thought was called Behaviourism. They argued that humans, like animals, are biological machines whose behaviour is no more than a conditioned reflex to external stimuli. Dismissed consciousness and the soul as figments of human imagination.'

'So they conceded the concept of imagination?' I asked, confused.

'I don't know that much about it. I *think* Skinner conceded that humans might have something like a psyche. But he thought that such a nebulous concept would be difficult to find in the human body and that remote control would do a better job of running it. *Beyond Freedom and Dignity* – that's what he entitled his thesis. You've read *Brave New World* though haven't you?'

Brave New World, which painted Aldous Huxley's futuristic vision of a society engineered by biochemists and pharmacologists and controlled by a powerful élite of security-seeking capitalists, was widely regarded as the most prescient of the dystopian novels of the 1930s, albeit less popular than Orwell's *Nineteen Eighty-Four*. Huxley came from an eminently intellectual family and his grandfather, Thomas Huxley, was one of the most highly esteemed scientists of his time. Aldous trained as a doctor before dedicating himself to a literary career. In his writings he sought to communicate scientific advances and their implications to the wider public. My father had instilled in me a respect for Huxley at an early age and, as Garth knew, I had most of his books at home.

'Yeah, a while ago. But it doesn't mention Behaviourism.'

'That's what it's all about,' he said in an exasperated tone. 'He refers to it as Pavlovian conditioning. The main bloody character is called John Watson.'

'Are you sure?'

'Yes! Remember in the beginning, the babies are trained to associate flowers with fear?'

'Yes, yes, but I think that's only one scene.'

'The whole thing was about a Pavlovian-conditioned society. He explains what he was trying to illustrate in *Brave New World Revisited*, which he wrote when he was really worried about the communists using it for mind control. So was the CIA. That's how they justified their LSD experiments.'

'Thanks Garth. I'll have a look.'

'Well,' he stood up to leave, 'I'm not sure it's of much relevance to what you are doing, really.'

'It might be.' I thought about the psychology experiment in which they claimed to have conditioned human subjects to associate stress with a particular scent, to the extent that they physically suffered on subconsciously detecting it.

At home, I sat at my table and looked for references to Pavlov in *Brave New World Revisited*.

> For the dictator and his policemen, Pavlov's findings have important practical implications. If the central nervous system of dogs can be broken down, so can the central nervous system of political prisoners.

Huxley wasn't referring to people behind bars when he talked about political prisoners.

> It is perfectly possible for a man to be out of prison, and yet not free – to be under no physical constraint and yet to be a psychological captive, compelled to think, feel and act as the representatives of the national state, or of some private interest within the nation, wants him to . . . The victim of mind-manipulation does not know that he is a victim. To him, the walls of his prison are invisible, and he believes himself to be free. That he is not free is apparent only to other people.

I looked up from the book and out of my window. It was dark outside but the neon lights of the supermarket across the road illuminated the faces of passers-by. I watched late-night shoppers' fearful reactions to a local madman who lumbered slowly along the street below me. I knew his movements well enough, from observing him on other occasions, to appreciate both his harmlessness and the fact that for anyone who hadn't come across him before he could be very scary. Over six foot tall and of a heavy build, he marched up and down the High Street, pausing to roar every twenty steps, regardless of who happened to be standing in front of him. His heavy movements reminded me of a schizophrenic client I'd had who told me he was being issued subliminal orders by the security services and attributed his actions to their external control. I thought about the theory that those diagnosed with schizophrenia were peculiarly sensitive to olfactory messages. And then I thought about the dogs on mental health wards. I'd managed to get hold of the responses to the consultation exercise conducted prior to the decision to use dogs in mental health hospitals. Patients and medical staff had expressed a number of objections to letting the dogs in; they said it would breed fear and exacerbate paranoia by putting them under permanent suspicion and talked about 'police induced psychosis'. One patient said that the proposal was evidence that the police officers were less sane than the patients; 'you can't have a conversation with a springer spaniel.' The breed of dog mattered greatly to some patients. Spaniels would be generally all right, but terriers and Alsatians would often not be. One member complained of an incident when they were told that a dog of a certain breed would be visiting, and when the day came it was another breed entirely. This was felt to be a betrayal of trust, even though it was explained that this was because the dog in question was all that was available on the day. It was mentioned that there was irony in such a heavy-handed crackdown on patients using drugs of their own choice given that entering hospital usually means patients are forced to take many drugs and in much larger doses than they would willingly do in order to keep them manageable. This comment reminded me of the Ronald Reagan look-alike's plans for scanning the brains of drug users to find out the cause of their disobedience. I felt a growing apprehension of neuro-regulators.

Back in Huxley's book, he quoted extensively from the internationally renowned social psychologist, psychoanalyst and humanistic philosopher, Erich Fromm. Fromm argued that Western civilisation was increasingly less conducive to mental health, reducing the individual to 'an automaton who pays for his human failure with increasing mental sickness'. According to Fromm it wasn't people like the man I could see out my window we had to worry about, but the apparently 'normal' sections of the population who failed to exhibit any mental health symptoms in response to their increasingly authoritarian environment. There was little hope for the normal, he said; they had stopped striving for happiness and given in to external control; 'man is not made to be an automaton, and if he becomes one, the basis for mental health is destroyed.' I looked back out the window at the man, who was now shouting, and wondered if he was rebelling or taking orders.

I tried to picture the 'normal people' Fromm was referring to and smiling work colleagues sprang to mind. I'd had a number of run-ins with management recently and was feeling increasingly suffocated by the atmosphere in the department. In management's opinion, I had a problem 'with the concept of employment'. 'What sane individual wouldn't?' I now asked myself. I'd always hated office workers who put signs up saying 'You've got to be mad to work here', but maybe they had a point.

Huxley blamed the Soviet adoption of their notorious police methods on the military realisation that Pavlovian methods could be applied to human beings. He called the Communist security services 'Pavlovian policemen'. He wrote *Brave New World Revisited* in 1959. At that time, the most prominent dictatorial power was that of the Communists in the Soviet Union. They were renowned for their secret police and the scientifically inspired inter-rogation techniques that they practised. The East German Stasi's use of scent sampling is a prime example.

I looked chemical warfare up in Huxley's *Encyclopaedia of Pacifism*. It had been published in 1937, shortly before the outbreak of the Second World War, when the world was worried that further conflict would see a return to the chemical and biological weapons that had been used in the First World War, when the technology behind them had been in its infancy. Although these weapons had been banned by the Geneva Convention, Huxley was well aware that despite the public's abhorrence of their use, a number of countries, including his own, were continuing to develop them. In his writings, he urged countries against this activity, reminding them that they represented a danger-ously anarchic and uncivilised form of warfare. 'During the Crimean War Lord Dundonald proposed that sulphur fumes should be used against Sebastopol. The War Office rejected the proposal on the grounds that "an operation of this nature would contravene the laws of civilised warfare".'

Sulphur, I knew, burns with a blue flame that emits sulphur dioxide, notable for its peculiar suffocating odour.

In the aftermath of the Second World War, Huxley found little comfort in the fact that chemical and biological weapons hadn't been used. The Allies' discovery of vast supplies of chemical weapons in abandoned German arma-ments factories, and the fact that the Soviets would now be in possession of some of them as a result of the post-war division of spoils, ensured that research into their devel-opment was continued by all sides. It was when reports started coming in about Soviet interrogations and police malpractice that Huxley switched his attention to the use of chemical weapons on domestic populations. The CIA was also perturbed by the possibility that the Communists were developing the power to control minds. Their solution was to start conducting their own research into this field. That was when they started experimenting with the effects of LSD without their targets' knowledge or consent.

Huxley wouldn't have been pleased to see the growing influence of arms companies in domestic affairs. He'd thought that their impact on foreign affairs was bad enough:

> The desire of arms manufacturers to make profits is a standing menace to world peace. It is in their interest to work for policies which are likely to produce dangerous international situations and to work against disarmament and the establishment of world peace, which would spoil their trade.

As Martin had suggested, it was high time for me to get stuck in to the literature on chemical and biological weapons.

I headed for the library first thing in the morning. Security guards had been hired to search readers' bags. I handed my bag over and exchanged smiles with the jolly-faced man as he unzipped it.

'Anything sharp in here?' he asked.

'No.'

He laughed. 'We should give people their own bags to search really shouldn't we?'

'There's no bombs in there either.' I smiled.

'Oooh. Don't use that word,' he said, using his pupils to point at his superior. 'A lot of boomaphobia going on in here.'

'Boomaphobia?'

'Yeah, boomaphobia – fear of getting blown up on public transport. I've convinced a lot of people it's a real condition.' He laughed.

'Well, maybe someone should diagnose it as a medical condition. It seems to be spreading.'

'Yeah, but I can't complain,' he said. 'It pays my bills.'

'Yes, I suppose it does.'

'I just have to phone up the BBC every two weeks and pretend I'm going to do something.'

'Really?' I said, slow to pick up on the joke he was making as my mind was already inside the books I wanted to look at.

'No, not really.' He handed me back my bag.

I was surprised by how much literature there was on Western interest in the development of chemical and biological weapons. From reports in the newspapers, I'd got the impression that it was something to which the poorer, 'uncivilised' countries resorted. Most of the research into it was being done in the name of less-than-lethal weapons and the applications being dreamt up were wide-ranging. In 2002 the director of the US National Institute of Justice (NIJ) stated in her summary of a NIJ report on aircraft security that:

> Anaesthetics or calmative chemicals could, in principle, be developed into a system whereby they could be remotely released into the cabin in order to incapacitate all passengers, and the hijackers, until the plane can be landed safely.

The extract reminded me of Roger's aeroplane vents. Was this where he had got the idea from? Why not just knock everyone out at the beginning of the journey as a security measure? They just wanted to get where they were going without noticing the journey. First class already looked like a hospital ward; equipment haphazardly wired to horizontal passengers. Of course, as the world had learnt from the Russians' use of less-than-lethal weapons, pumping them through the ventilation system of the Moscow auditorium in 2002 in which some 800 people had been taken hostage by persons claiming allegiance to the Chechen separatist movement, the aeroplane vent plan wasn't as straightforward as it sounded and could well result in the safe landing of a number of corpses.

A joint Russian–American conference established among the desired functions of non-lethals the ability to 'neutralise mass demonstrations' with 'disabling chemical compounds'. These could 'temporarily immobilise the mental faculties' thereby 'neutralising a person's will to resist'. Alternatively, they could 'produce physical distress for a significant period, such as severe discomfort, anaesthesia, paralysis or immobility'.

There were other proposals for agents that keep a person awake, but without the 'will or ability' to carry out criminal activity.

The principal challenge faced by less-than-lethal lovers is that individuals often come in groups, sometimes deliberately so. Different people will react differently to the same dose of weapon, as illustrated by the Russians' theatrical gassing. It is this reason that is given for the growing interest in the development of non-lethal techniques with a high degree of specificity. Present day less-lethal-weapons proponents claim that they have, or at least could soon have sufficient knowledge to target an individual and calculate the agent's impact with precision. I couldn't see how they could do that without our medical records as people vary in their ability to withstand the sorts of assaults these people had in mind. The information provided by biometrics could be used to feed the desire for such capabilities in less-than-lethal technology. In relation to 'specificity of wounding' two Chinese authors in the 2005 *Military Review* claim 'if we acquire a target's genome and proteome information, including those of ethnic groups or individuals, we could design a vulnerating agent that attacks only key enemies without doing any harm to ordinary people'. While it occurred to me that these people, like those at Olympia, were probably just trying to sell the technology of tomorrow today to technophiles in government, the fact that the desirability of such weapons was being discussed perturbed me. Some proponents were so excited about the growing acceptability of less-than-lethal weapons they were beginning to view it as 'a great commercial opportunity, which can later extend to the entire ammunition market'.

According to the British Medical Association (BMA), there is at present a 'widespread interest expressed by governments in the use of drugs as weapons'. The BMA views the government's interest as 'dangerous' and capable of including 'intentional manipulation of people's emotions, memories, immune responses, or even fertility'. The BMA identifies the driving force behind this growing interest in the use of drugs as proponents of less-than-lethal weapons for law enforcement. It claims that developments in this field are driving a horse and cart through the Chemical and Biological Weapons Conventions. This was interesting. I was beginning to agree with the international security expert from the Swiss conference that Orwell's *Nineteen Eighty-Four* had 'long since passed'. Perhaps we had entered the second stage; technologies of detection in place, the emphasis was switching to control as we were rapidly sucked into Huxley's *Brave New World*.

Sniff fast to be a 'human bloodhound'

Metro
18 December 2006

Researchers claim that, just like dogs, humans can follow scent trails across the countryside by using their sense of smell alone.

. . . A team laid a pungent trail of essential oils, containing chocolate, in a grass field and asked 32 people if they could follow the 10m-long line by using their noses. Two-thirds were able to do so . . . It was found that they got round trails faster as they became more adept at detecting scents . . .

CHAPTER TWENTY THREE

In the Name of the Nose

Our grasp of reality owes almost as much to politics as it does to physiology.
    Lyall Watson, *Jacobson's Organ and the Remarkable Sense of Smell*

When Ben Teckler informed me of a workshop on smell he was attending, I made it a date. I thought there might be a link between Huxley's concern with mind control and Pavlovian conditioning, the growing interest in using drugs as weapons, and Teckler's statement that neuroscience was the most relevant scientific discipline in olfactory research. He told me that security was unlikely to be on the agenda as it was an informal grouping of scientists working on olfaction. But if the MoD was interested in what these people had to say, then so was I.

Two weeks later, I was in the countryside, rubbing shoulders with the country's leading olfaction experts, watching an MoD scientist exchange pleasantries with a perfumer from the Cotswolds over coffee and chocolate biscuits. I hadn't appreciated how powerful the fragrance industry was until I'd chanced upon an article, *The Scent of Fear*, in the anti-consumerist magazine *Adbusters*. It named six companies – Firmenich, Symrise, Givaudan, Quest, Takasago and International Flavors and Fragrances (IFF) – as being responsible for the ingredients of over half of what we smell, 'from the unglamorous functional perfumery of hairspray, toilet bowl cleaner and underarm deodorant, all the way up to the prestigious realms of fine perfumes and colognes'. Fragrance was a multi-billion pound industry. 'Why have I wasted all this time worrying about arms companies?' I asked myself. I'd heard it said that life in the future would be like living in shopping 'malls without walls' but I'd thought it was a metaphor. In fact, it was beginning to look like a descriptive premonition.

I'd been nervous about seeing Ben again. We'd shared several friendly email exchanges but hadn't seen each other since the MoD conference. But he was as easygoing, kind and helpful as he had been all along. He introduced me to people when he could, filling me in on their background and on the various areas of contention in the olfactory sciences. The main source seemed to be the divide between 'the shapists and the vibrationists'. This, Ben explained, was an argument over whether there was one olfactory receptor for each molecule – which would mean that each receptor only registered the odour if it was the right fit – or whether each receptor was capable of recognising many smells and measured them by their respective vibrations. Most people here were shapists and were displeased with the popularity of Luca Turin's *Emperor of Scent*, in which he extols the vibrationist theory and accuses shapists of sabotaging his research. The topic was hotly debated over the conference meals. I was surprised to find out that no one knew what made a molecule smell the way it did, and to hear that chemically similar molecules could smell very different. The olfactory code, it seemed, was yet to be cracked.

Only a few of those in attendance were from the perfume industry but only one other woman, Hilda, was like me a novice in the world of smell. She had undertaken a perfumery course with one of the group's members and become fascinated with the topic as a result. Most people there were academic scientists. Dick Doty was probably the biggest name. He had edited the *Handbook on Olfaction*, among other publications, and was a leading world expert on the human sense of smell. Ben recommended I talk to him. Unfortunately, that's what most people there wanted to do.

The sense of smell had been belittled by the mainstream scientific community for several decades and this was a source of great bitterness among the olfactory community I was now befriending. They were of the belief that the importance of the human sense of smell had been gravely underestimated. 'After all,' cooed the chairwoman, 'if you were taken into a supermarket with your eyes closed – how would you know where you were?' 'The secret of the horse whisperer,' chimed in another member, 'is his use of scents to influence the animal's behaviour.'

Out there on the sensory battlefield, however, the tide was finally turning. It was predicted that, in the not-too-distant future, the amount of knowledge about the olfactory system would rival the level of knowledge in the visual sciences. This, it was believed, would have far-reaching consequences, as suggested by the late Lewis Thomas, physician, essayist and policy advisor:

I should think that we might fairly gauge the future of biological science, centuries ahead, by estimating the time it will take to reach a complete, comprehensive understanding of odor. It may not seem a profound enough problem to dominate all the life sciences, but it contains, piece by piece, all the mysteries.

A sign of this tempestuous change in the sensory season was that we were to be honoured with a presentation from an eminent neuroscientist who had joined our workshop. The neuroscientist explained that research into the neural processes involved in odour perception had moved away from rats and was focusing increasingly on the human. Advances in neural-imaging technology permitted new insights into olfaction. He was currently engaged in Pavlovian conditioning experiments on humans, using odours as the conditioning stimuli.

An academic expert on the psychology of smell, and long-standing member of the group, ran through some recent olfactory highlights. Top of the list was the invention of the smell cannon, capable of tracking an individual and firing scent rings up the target's nose from a distance of up to five metres, leaving the person next to them completely unaffected. The device tracks the person it is aiming at with a camera mounted on top of it. Software on a PC analyses the video images and controls the motors steering the gun. Once it has a fix on the eyes, it aims low to direct the puff of air. When the cannon is fired, a fine jet of aroma-rich air is forced in the required direction.

He pointed out that attractive scents, like the smell of freshly baked bread, are already known to keep customers in a store for longer, but retail areas, until now, could only be infused with one odour at a time. The air cannon would allow different scents to be fired at individuals. This could have major commercial applications for individualised advertising. He warned that there might be civil rights problems with using scents in this way, however, as customers might object to having scents forced up their noses.

A young woman chemist from Quest explained how odours travelled directly to the limbic system, provoking powerful emotions. Quest and other perfume houses are now using neural imaging techniques to measure the effects of their scents. Odours are of great value in a wide range of products other than perfume. You can 'kid people' about the texture and attractiveness of a product with odours. I thought about the bacon roll I'd been tempted to buy on the train on the way up. A man across the aisle from me had opened up a packet with one in and a cursory glance had informed me it wouldn't be as good as it smelled.

The Quest woman took a collection of vials out of her apparently bottomless bag and distributed them among us. 'Androstenol,' she said, 'is interesting. Only some of you will be able to detect its smell.' When a vial arrived I took a whiff. I couldn't smell anything. Were my phero-mone detectors defunct? Surely not. I remembered how Tom's breath smelt back at my house, where he'd carried me on our first date. I'd been amazed to find him in my bed on regaining consciousness. I'd sat up and turned my back on him to pull off my dress. I jumped down from the raised bed and pulled a nightie out of the cupboard. After putting it on, I quickly brushed my teeth and jumped back into bed and under the duvet, tousling his hair before turning my back on him to return to sleep. Our feet quickly found each other under the bedclothes. With each new and potentially accidental point of contact layers of sensation seemed to build up inside me, spreading out further and further until my body was almost touching his and then the palms and fingers of his hands stroked my hips and I lifted my face up, losing my breath on finding his. I wanted to climb right inside it and could feel myself dissolving as his scent filled my lungs.

Then he pulled back. Said 'we shouldn't.' Flabbergasted and suddenly conscious of my drunkenness, I mumbled a barely audible objection. It was impossible to describe the smell of his mouth but it was warm and kind and welcoming. It made his sudden detachment painful. He took me under his arm and held me there. A couple of hours later he left me to lounge in the smell he'd impregnated my sheets and pillows with. I buried myself in them for days, dreaming of what should have happened next.

'Although only some of you can consciously smell it, all of you are registering it and your brains will have responded directly to the stimulus.'

'Could you smell it?' asked Hilda, who was sitting next to me, disgust written all over her face.

'No. What does it smell like?' I asked.

'Horrible. Sniffing it was like having my head dunked in a urinal.'

The *Adbusters* article had described the six major players in the fragrance industry as the 'ghost writers of the olfactory universe'. And here was one of their chemists explaining how she was trying to tap into 'Proustian fragrance'. Roger's nightlight blinked in the distance.

The table in the waiting area outside the seminar room, where the coffee and biscuits were supplied, was covered with back issues of *The Aroma-chology Review* and other fragrance-oriented literature. It made for interesting reading. The olfactory science community seemed to be on a high. They were giving each other awards and instituting national awareness days. They said a sensory revolution was taking place and smell was nearly in the lead. Mainstream science has started to recognise the relevance of olfaction, they said, after years of refusing to acknowledge it, and on some occasions, even ridiculing it. Psychophysiology, for example – the study of the relationship between physiological responses such as brain activity or heart rate and mental phenomena such as emotion or thought processes – had recently caught on to the potential insights the study of smell could provide. As had neuroscience.

The fragrance industry, indeed the olfactory community as a whole, was benefiting from the expertise that was being brought into their field. The human olfactory system, it had now been discovered, was remarkably similar to that of the rest of the animal kingdom. As well as projecting directly into the limbic area of the brain, smells feed the thalamus and neo-cortex in the 'new brain', where they can be consciously perceived. For some reason, olfactory signals induce emotional reactions regardless of whether they are consciously perceived, and odours, it is said, have more potential than any other social environmental sensory stimuli to influence human behaviour. While the distinction between the conscious and the subconscious remains in many respects a philosophical one, recent research suggests that the subconscious, far from playing second fiddle to the conscious mind, is orchestrating a wide range of the mental faculties we regard as uniquely human, like imagination and memory. Whereas conscious thought processes are disrupted if you are forced to direct your attention elsewhere, subconscious ones are not.

'Environmental aromas', the concept of incorporating scent into the architectural design process, was evolving rapidly as a result of collaborations between the fragrance industry and 'modern ventilation systems'. Plans were being made for 'a healthier, more refreshing experience indoors for everyone'. In the perfume industry, it was a well-known fact that in many instances 'odours have to be hidden to exert their effects'. Whereas the practical utility of fragrance systems was held back in the past by the difficulty of controlling its dispersion, transmission through the air is now controllable for the first time. It was clear from *The Aroma-chology Review* that the perfume industry was keeping a hawkish eye on the Food and Drug Administration (FDA) in the US, and gearing up for battle with it. The FDA had announced it was considering re-classifying aromatherapy products as drugs instead of cosmetics:

> The FDA's position underscores the importance for the industry to consider adopting the term 'aroma-chology®' which was coined by the Olfactory Research Fund. As aroma-chology is concerned specifically with the temporary, beneficial psychological effects of aromas on human behavior and emotions to improve mood and enhance quality of life, it does not fall under the FDA's definition of 'therapy'.

A more explicit security agenda crept into some of the articles. Scientists were asking whether aggressive or fearful people might be identifiable by their smell and researchers undertaking Pavlovian training of humans using odours were experimenting with peculiarly negative emotions. Under the strapline 'the smell of failure', researchers from Monell Chemical Senses Center recounted how children's cognitive ability could be dampened by exposing them to an odour they'd been conditioned to associate with the feeling of failure. The conditioning was straightforward. A child was placed in the room and presented with an unsolvable maze. The child was told that if they could solve the maze, they would be given a cuddly toy. While struggling with this impossible task, odours were delivered into the room using a 'hidden environmental-odor delivery system'. Shortly after failing this first test, they were relaxed by being read a short story and then taken into another room to perform a cognitive test. Once again, the same odour was released into the room. The children who performed the cognitive test in the presence of the same odour did worse than the unconditioned children. The results suggested that the neutral odours used had 'acquired specific, emotional associations' in the children's minds that could alter their behaviour and performance. The experiment was reported to have 'important implications for . . . any setting where ambient odor scenting can be achieved'.

The Monell Chemical Senses Center had begun examining the psychological characteristics of an individual that may contribute to the physical symptoms experienced on exposure to certain odours. Results suggest that people who worry about environmental exposure may be predisposed to experience physical symptoms including headaches, eye and throat irritation in response to smelling a usual odour: 'Since personality characteristics appear to play a major role in understanding people's response to olfactory stimuli, it is important to realize that subjective health symptoms that are elicited from exposure to odorous substances may not be prevented by a simple focus on the properties of the chemicals themselves.'

After the coffee break, a chemist with eclectic research interests gave a talk on how cockroaches 'instinctively seek out cracks in walls through which they can dangle their antennae in search of interesting odours whilst protecting their bodies from human enemies'. He explained how electrodes could be implanted in the antennae to see what compounds the beetles are detecting. No mention was made of potential security applications but at least one sprang to mind. Walls wouldn't just have ears in the future; they'd be fully sentient organisms.

Mosquitoes find us by detecting the volatiles (lively molecules that can fly is how I thought of them) that emanate from our skin and sweat. Perhaps, I thought, it was only a matter of time before mosquitoes would be used to harass enemies of the state apparatus. Scientists had discovered that human sperm contained olfactory receptor proteins, which it used to successfully locate ovaries. No suggestions were made to re-wire its target or weaponise it but the possibility swam menacingly across my horizon.

The research on Jacobson's Organ suggested its connections to the human brain had been unplugged. However, research on pigs showed that the main olfactory system was capable of picking up on the pheromonal messages previously thought perceptible only by Jacobson's Organ. One whiff of androstenone, for instance, and 'a sow will immediately assume a mating position'. It was possible that buried deep down within the structures of the human psyche lay ancient patterns of reactions to certain environmental stimuli that could still be provoked despite the centuries of neglect caused by their redundancy in a modern environment.

A representative from a company specialising in marketing with scent told us of the intimate connections between the olfactory system and the emotional centre of the brain called the limbic area. Of all of our human senses, she said, the olfactory sense has the greatest and most direct influence on our emotions. Not only might it have a more immediate and deeply penetrative impact on our minds than the visual sense, but it promised other advantages in terms of advertising as well. Visual advertising, she said, is no longer as effective as it once was because people are wary of such adverts and tend to ignore them. 'Screening senses' like sight and sound appeal to our rational mind. 'Trusting senses' like touch, smell and taste appeal to the emotional brain – 'we believe them'.

Their effect is immediate and subliminal. She said that we always believe what our nose tells us. I thought about Lyall Watson's description of the vulnerability of the olfactory sense. 'Most other sense organs lurk under the skin, embedded deep in protective tissue, picking up sensations by remote control. But smell cells go unadorned: naked neurons, each one right out there in the open, like a unicellular organism, meeting molecules, making its own way in the world.'

I looked up from my notebook and smiled across the room at Ben. It was good of him to let me in on this conference.

More and more shops, the representative enthused, are cottoning on to the benefits of adding specially designed fragrance into the air-conditioning systems. Odour is an effective means of securing a 'lasting relationship' between the brand and the consumer. Whilst organisations like banks were experimenting with devices like scented cheque-books, the future will be about 'three-dimensional branding' in which companies can purchase airspace. I asked the speaker if odour presented companies with a means of subliminal advertising that it would take regulators a while to catch up with. She smiled proudly and said that it did.

The cancer detection dog crew were next up and they were full of beans. They'd just published their research in the *British Medical Journal*. They'd had 'a heck of a time' trying to convince the medical establishment that they were on to something. Did we want to see one of their dogs in action? 'Yes!' we cried. It was like being offered a magic trick. We watched as the handler laid the pots of urine across the floor in front of us. Did we want to know which one was cancerous? 'Yes!' She told us quietly which one it was and then the dog was brought in. The dog ran *straight* for the cancerous urine and sat down. It was a miracle!

We broke into rapturous applause. Did we want to see it again? 'Yes!' The dog was taken out and fresh samples laid out. Did we want to know which one was cancerous? 'Yes!' The dog was brought in and ran *straight* for the wrong pot. We looked up at the handler. She was frowning. 'Ah,' she said. 'I think that sample was from a menstruating woman. This dog loves menstruating women.'

The cancer detection scientists were upfront about the enormous hurdles they faced. Although there was a large body of anecdotal evidence of people's pets barking at cancerous growths, it wasn't yet clear if cancer had a particular scent, or if dogs were picking up on the scent of changes or weaknesses in the immune systems of their owners rather than the cancer itself. The dogs were good at picking up on the scent of disease in the urine pots but it was difficult to know what they were training them to detect, because they didn't know which scents out of the generous bouquet supplied by a urine sample the dogs were actually picking up on. These hurdles, they were convinced, could be overcome in time, but they were finding fund-raising very difficult. It wasn't that nobody wanted to know whether they had cancer. Everyone wants to know that. It was just that the medical establishment refused to believe dogs would ever be capable of correctly diagnosing it. The rigour of scientific methods bit when dogs were proposed as a diagnosis tool for disease, unlike in the forensic field, where they apparently lacked teeth. A number of hospitals were averse to allowing cancer detection dogs within their establishments, on the basis that they were unhygienic. I'd not yet heard of a hospital refusing entry to drug detection dogs on this ground.

Over lunch, I discussed the positive regard in which drug dog indications were held compared to the cancer detection dogs. An elderly female scientist, who had worked with both police dogs and cancer detection dogs, explained how much stricter the controls used in the cancer detection experiments were. Police dog handlers, she said, have a vested interest in seeing their dogs do well – and this bias affects the accuracy of field tests. I got talking to one of the cancer detection dog handlers. She told me her dog would often show an interest in strangers they came across on their walks, which she felt put her in an ethically difficult position; should she tell them they had cancer? I was unsure how many pedestrians would have been perturbed if she had, though I noticed I'd been keeping an eye on the dog to see if it showed any interest in my odours. I am a hypochondriac at heart and it was a good thing she didn't have any dogs for sale there or I'd have ended up with one, just as a security measure.

Ben sat opposite me at the lunch table. I told him about some of the novels I'd been reading that talked about smell.

'I find them easier to read than most of the academic literature!'

'Of course.' He smiled. 'I hadn't realised that there were many novels on smell. You must tell me what you've read.'

'Well, I found this great sci-fi novel from the 1930s called *The New Pleasure*. It's about a scientist who invents a drug that heightens the olfactory sensitivity of humans, with the result that people start to take better care of the environment. It's by John Gloag.'

'Well, he wasn't that far off in his imaginations. The US government had a project for making such a drug.'

'Did they manage it?'

'No, but they poured vast amounts of money into the project. Of course they also tried planting mammalian olfactory cells in soldiers' noses and that didn't work either. We aren't allowed to do that sort of research in the UK.'

After lunch, Professor Richard Doty tore the literature

on pheromones to pieces in his presentation, demonstrating step by step why most of what was written about them was nonsense. Apart from the definitional vagaries, which he illustrated with humorous aplomb, he argued that there was no such thing as an innate pheromonal reaction to external stimuli. Nature had provided all creatures with a flexible sensory system capable of categorising novel agents. It might be that organisms were born with a subtle disposition to certain chemicals but essentially, our sensory system learns the meanings of odours through experience. 'Learning,' he said, 'overrides chemosensory genetics.' Moths could be fooled into tracking down marijuana instead of female mates. This demonstrated the plasticity of biological organisms.

Richard Doty turned out to be highly charismatic and quick-witted and after the talks were over he entertained us all in the pub with anecdotes from his different walks of life: gambling, car racing, and his favourite claim to fame, which was that *Hustler* magazine had described him as 'sick' for his research on the odours of vaginal secretions. 'Not many people get called sick by *Hustler*!' Most of the scientists were good fun in the pub. No one from the perfume industry joined us and everyone seemed to be on friendly terms. The psychologist appeared initially unnerved when he found out where Ben worked, but it was difficult not to like Ben and we all passed the evening happily together. A long-standing member of the workshop told us of a phone call he'd received from Her Majesty's Customs and Excise. They were now paying him to find a means of impregnating banknotes with an odour to facilitate their detection. 'Now I've got a licence to print money!' he smiled gleefully, showing us a copy of it. Ben said he had stacks of material on odour conditioning in his office, which he'd try and dig out for me. A bird olfaction enthusiast told us vultures were being used by the Belgian authorities to locate decomposing bodies in a forensic revision of the ancient practice of falconry. No doubt Ben was right about this group. Security wasn't their main interest. But its presence, like in so many other walks of life, was increasingly pervasive.

Doty and I sat alone in the hotel bar. Before saying goodbye, Ben had given us all a lift back, letting us poke fun at him for not driving a James Bond car with the impressive gadgetry we felt befitted his Q-like role in the defence of the realm. The rest of the people who had joined us in the pub went straight to bed when we got to the hotel. Neither Doty nor I was tired enough to follow suit.

Over the course of several whiskies, Doty told me about how he'd been roped into providing expert evidence at a criminal trial in the US to challenge a statement by a police officer that he could smell the defendant's cannabis plants from a road several hundred yards away. What had happened, Doty told me, was that the defendant, who was cultivating cannabis in his home, and was 'a very nice guy actually', had an argument with his girlfriend. The girlfriend reported his cannabis cultivation to the police but retracted her statement before they visited his premises. The police therefore found themselves without any lawful reason for busting down his door. And that was where their sense of smell came in handy. The police officers said they could smell the plants, and broke in on that basis. On behalf of the defence, Doty conducted the first empirical experiments on the ability of humans to detect the scent of marijuana plants at a distance. His research made a mockery of the police officer's claims.

The perfumer from the Cotswolds kindly offered me a lift to the station the next morning. His claim to fame was having been commissioned to make personalised perfumes for the Queen and other members of the royal family, a mission he had completed successfully. On the way to the station, we talked over the presentations we'd heard and he filled me in on the background of some of the people there. I'd been particularly intrigued by the lady who had worked with police dogs and cancer detection dogs, having established that she'd done some work for criminologists.

'Yes, well you know one of her big research interests used to be the scent of fear.'

'Really?' I said, suddenly scared of her. I wondered what effect that scent would have on humans and heard the snapping jaws of my reptilian brain threaten to grind up my thought processes.

'Yes, in the 1980s when she was looking into human scent. She wanted to take scent samples from arachnophobes surrounded by spiders and the like. Her research was beset with ethical obstructions though, and so she ended up taking up my suggestion of accompanying first-time parachutists.'

'Did she get anywhere with it?'

'No. I'm not sure what happened.'

I thought back to the Swiss conference and the participants' desire for sensors that could monitor physiological responses in order to detect nervous visitors to public areas. Olfactory versions of lie detection tests were easy to envision. Providers of neuro-imaging devices were already translating our cognitive processes into coloured pools of ink and claiming that they could piece them together into pictures of guilt. Nosey interrogators, I feared, would get their wicked way eventually.

Could the smell of fresh bread kill?

Metro

2 February 2007

BAD SMELL: The smell of freshly baked bread might help sell your home but it could make you die younger. Hungry fruit flies exposed to the smell of yeast do not live as long as other flies on similar diets but not exposed to the odour, researchers found.

Flies with a genetically reduced sense of smell also lived longer, the US team reported. They think the smell of food interferes with animals' ability to cope when there is little food about.

CHAPTER TWENTY FOUR

The Malodorous Twilight

The policemen pushed him out of the way and got on with their work. Three men with spraying machines buckled to their shoulders pumped thick clouds of soma vapour into the air.
   Aldous Huxley, *Brave New World*

The more I read about the powers of smell, the more wary I grew of the fragrance industry. Romantic visions of pretty glass bottles and lavender fields caved in to vast underground networks of chemical laboratories as my mind forged synaptic connections between weapons of mass destruction and scents of slow seduction. The scent of fear, I'd discovered when browsing through a list of pseudo scents for police dog trainers, had already been bottled. One contributor to the K-9 Academy for Law Enforcement Trainer Resource Centers web page suggests that 'if you want to train your canine to pick out all the people who are afraid, you can do that . . . a guilty person's odor is coming out like a smoke bomb and smells of fear'. I worried that this smell might be used to induce fear in enemies of the state in a sophisticated version of the bee-keepers' smoke machine.

'What do you mean?' Cass stopped me when I confessed my fragrant fears to her over the phone. I'd started by telling her about how the synthetic ingredients for the fragrance of musk were discovered during work on explosives.

'Yeah, they both involve chemistry. Big fuckin' deal.'

'Most famous fragrance houses were taken over by pharmaceutical companies in the 1960s.'

'OK, I don't like pharmaceutical companies. They are the bad guys, but c'mon, chicken. Get back on the ball.'

'And the Joint Non Lethal Weapons Directorate in the States, and the UK government, have both identified mal-odorants as one of the most promising technologies for the future needs of law enforcement. It's being argued that they are permissible under the Chemical Weapons Convention.'

'Stink bombs?'

'You could call them that if you wanted to trivialise the issue.'

'Well excuse me Marjorie!'

'The use of malodorants has a long history. There has been a renewed interest in it recently. The Sunshine Project . . .'

'The guys who campaign against biological warfare?'

'Yes.'

'They are working on stink bombs? I mean malodor-ants?'

'They are particularly worried about them. The Chem-ical Weapons Convention bans munitions and devices for their dispersal and government contractors are now making them and saying they're for malodorants.'

'Like what?'

'Microencapsulation, ventilation ducts, air dispersal units, gun-like payload dispersal systems, multi-sensory grenades. And then there's the Biological Weapons Convention. Most offensive smells are produced by living organisms or are toxins derived from them.'

'Right, so they're not worried about malodorants per se. They're worried about them being used as a cover for developing other weapons and *calling* them stink bombs?'

'Well, yes, I suppose that could be a problem. By definition, an odorant is a chemical that induces the psychological perception, which is termed odour, so any molecule that induces an odour is an odorant. And the olfactory nerve *can* serve as a conduit for the movement of viruses from the nasal cavity into the brain. In fact, it's a direct route. And mustard gas has a smell so I guess you could call it a malodorant. But I don't think they'd be that brazen in their use of terminology. In fact no one has suggested that they would be. The Sunshine Project is worried about malodorants whose sole impact can be attributed to the odour itself.'

'Honey, I think you might be missing the point. They're not worried about perfumes.'

'What makes you think that's such a joke? Anti-weapon campaigners suggest that all non-consensual manipulation of human physiology should be outlawed.'

'Well, these malodorants that they are researching, they don't hurt anyone do they?' I could almost picture her shaking her head in concern for me, her small, freckled nose sitting prettily above her thin, pink lips.

'Well, they are a *sensory irritant*. That is their key function. Official documents *claim* that their criteria for legitimate malodorants is that they *may not be* a sensory irritant, but like the Sunshine Project states, any such malodorant would be a contradiction in terms. Malodorants are designed to stimulate your olfactory receptors. A lot of them are designed to make you feel sick. They call them gastrointestinal convulsives. They do have an impact on the nervous system. Their intended application is the modification of human behaviour.

And I mean, OK, let's say they're not chemical weapons. Let's take Pamela Dalton's line. She seems to be doing most of the malodorant research for the military from her base at the Monell Chemical Senses Center. She denies that their use amounts to chemical warfare. She says it's more like psychological warfare.'

'I don't get it. How?'

'She's been experimenting with different odours, trying to find the most offensive one. She says the combined effect of her two most offensive concoctions, Stench Soup and Bathroom Odor, is so distasteful that it chases all thoughts out of the mind.'

Cass laughed. 'How do they want to use them?'

'To disperse rioters, get people to abandon plans to storm an embassy, induce mass panic. Back in the 1960s DARPA was talking about conditioning people to react in certain ways to specific odours. They gave an example of a bomb emitting an odour on explosion so that from then on they could rely on the odour alone as a fear-producing stimulus because of its limbic associations with the fear experienced during the bombing.'

'OK. That's fucked up.'

'The Sunshine Project has even uncovered ethnically targeted malodorants. Back in the 1960s, DARPA commissioned researchers to compile an odour index of cultural smell phobias. And the Sunshine Project is concerned that the resurgence of military interest in malodorants presents a danger that some might pick up where DARPA's work left off.'

'Doesn't that sound like what Roger was telling you?'

'Oh, yeah.' I'd forgotten about that. Dark shadows arched their way round his nightlight, dancing wildly.

'So maybe you are on to something with these smell weapons. Shit, what do I know?'

'Well, I've found a professor of German literature who's

done a thesis on Huxley's *Brave New World* being a future society artificially controlled with synthetic fragrances.'

'Killer! Have you talked to him?'

'Not yet. He's based in the States but he's visiting Germany next week and so am I.'

'You are?'

'Yes.'

'Will you see Werner?'

Werner Pieper, or 'the information shaman of High-delberg' as Timothy Leary had called him, was a German psychedelia-clad hippy-storyteller in his late fifties. He claimed that life was safe when you dressed like a clown. 'No one is going to mug me in these pants,' I'd heard him say to someone asking him how he'd found LA south central. He was a writer and publisher of sex, drugs, politics, esoterica and music books, and underground magazines. He was an old friend of Cass's and a new friend of mine. He'd been sending me bizarre articles on subjects he thought related to my research since Cass had told him about it, over a year ago. He lived outside Heidelberg in a house he'd invited me to visit.

'Yes. It'll be full moon while I'm there.'

For almost thirty years, Werner has made his home an open-house to friends at the full moon. He never knows who or how many will turn up, and his wife Nadina always prepares enough food for any number. Only once, Cass had told me, no one showed up. 'But they might not show until the middle of the night.' Sometimes 'some crazy dude in a star studded blue cloak'll turn up in the garden' or a guest will wander out into the woods behind the house and 'take acid on the hilltop with his dog'. I was looking forward to visiting Werner on his home-ground, having only ever met him at Cass's. I was only sorry that I wouldn't have time to stay long at his house, minimising my chances of meeting cloaked gentlemen and their psychedelic dogs.

Make mine a pint of Chanel, landlord

The Sunday Times
Robert Boothe
5 August 2007

Pubs are planning to pump in artificial scents to mask the smell of stale beer, sweat and drains that used to be disguised by cigarettes before the smoking ban.

. . . Supporters of the smoking ban insisted that pubs and bars would become sweeter smelling without cigarettes. But the smoke had masked the locker room aroma in some crowded venues on warm Friday and Saturday nights.

Oliver Devine, senior marketing manager at the Sizzling Pub Company, part of M&B, said: 'Appetising food smells have increased but others are less attractive, such as stale food and beer, damp, sweat and body odour, drains and - how do you put this nicely? - flatulence.'

. . . Jeff Mariola, president of Rentokil Initial's ambience division, which is supplying Marriott hotels with bar perfumes, suggested that the aroma of malted hops could encourage beer drinking and mask the smell of drinks trodden into the floor.

Luminar Leisure, which owns 122 nightclubs across Britain, has already started pumping a scent over its dance floors.

CHAPTER TWENTY FIVE

The New World Odour

There is an army of diseases under the Demon Pestilence stalking among us ... The vanguard of that army is invariably Stink.

Charles Henry Piesse, *Olfactics and the Physical Senses*

I had a number of reasons for going to Germany. A less-than-lethal weapons symposium was only the official one. Professor Hans Rindisbacher was willing to meet with me to discuss the importance of smell in Aldous Huxley's *Brave New World*. He said my research was timely and he'd be glad to discuss his own conclusions. 'I am sure you have heard about the former East German Stasi (secret police),' his email read. 'If you don't mind I'll take you for a little excursion in Berlin.'

A copy of Gustav Jaeger's *Origins of the Soul* had arrived at Werner Pieper's house. He'd had a hard time tracking it down and I was worried in case the book turned out to be irrelevant to the Dogwatch project, particularly since I'd learned that Jaeger's name was synonymous with woollen underwear, which Jaeger had designed for hygienic purposes in the nineteenth century. I'd seen Jaeger's *Origins of the Soul* referred to in a paper on the history of smell in literature but no one else seemed to have heard of it. Jaeger, a zoologist and contemporary of Charles Darwin, reportedly reckoned that scent was the seat of the soul. I hoped his thesis might explain my instinctive distrust of olfactory surveillance.

On the short flight from Stansted to Berlin, I read an article in the *New York Times* about how Soviet interrogation techniques formed the basis for the new style of 'enemy combatant' interrogations in the United States. A psychiatrist at Yale University asked: 'How did something we used as an example of what an unethical government would do become something we do?'

Outside Hackescher Market Station, I sat on the steps, looking out for what I imagined a professor of German literature might look like. The process was an awkward one. Most of the 40-year-old men I made eye contact with returned my gaze as if they too were looking for someone I might be. When Hans turned up, he didn't look like any of the men I'd been eyeing up. I'd made a clumsy mental picture of him from his curriculum vitae. Of Swiss origin, he was educated at the University of Bern and Stanford University and was now professor of German studies at Pomona College, California. He had wide-ranging literary, historical and philosophical interests and the olfactory publications stood out intriguingly amongst them. He'd published a book called *The Smell of Books* since his presentation on *Brave New World* at the Aldous Huxley Centenary Symposium and even taught a post-graduate course on smell in literature. I thought he'd be wearing glasses and a beige corduroy jacket with dark brown elbow patches on the jacket sleeves. But the man standing in front of me was far too funky for such garb. He was fashionably dressed, LA. Black and cream-striped trousers hugged his long, toned legs and his eyes sparkled with enthusiasm for life. Though his hair was grey, his face shone with youthful exuberance.

He greeted me with a palm pressed against each of my upper arms and a kiss on both cheeks. 'So nice to meet you Amber,' he said, stepping back to hold me at arm's length while smiling warmly into my face. I'd not expected him to have a Californian accent.

'Come,' he said softly. Waving his left arm in the air, he stretched his right leg out in front of us. 'Let's take a walk.'

'It's always nice to meet someone who shares an interest in smell,' I said shyly, slightly overwhelmed by his zest and not sure how else to start.

'For me too. Particularly someone whose research is not strictly chemical, behavioural or dealing with odour abatement! What you are doing sounds very interesting and timely.'

I followed Hans through the market and on to the main road.

'So you've heard of the Stasi?' he asked.

'Yes! They're one of the things that got me interested in olfactory policing.'

'Good, well, if you don't mind, I thought we could visit their former headquarters here in Berlin. They have turned it into a museum.'

'Wonderful.'

As we walked down the Frankfurter Allee, Hans pointed out the Soviet architecture. Huge monolithic apartment blocks lined the streets, 'the monumental classicism of the Stalin years'.

'You know that the police here in Germany have been accused of reviving Stasi methods? It was on the radio this morning,' he said as we strode down the pavement together.

'Really?'

'Yes, they've been collecting human scents to trace anti-globalisation activists they believe may try to disrupt the G8 summit in June. They have been confiscating sweaty vests and getting well-known radicals to impregnate steel pipes with their hand odour to present to their German shepherds.'

'What's the public reaction been?'

'Outrage, because of the Stasi connotations. And of course the German shepherd dog is synonymous with Nazis here in Germany.'

'Yes, I was wondering about the relationship between dogs and Nazis – do you know much about it? A German friend told me that SS officers had to fight a Doberman before being recruited.'

'I don't know about that. Adolf Hitler's faithful companion, Blondi, a German shepherd, was with him in the Reichstag bunker in Berlin. He killed himself after poisoning Blondi. Certainly the German shepherd is welded in the collective memory of the Nazis. The good thing about Germany today is its sensitivity to fascist politics.' I'd noticed this on visits I'd made to see my friend Käte in South Germany. Most of her friends, very few of whom could be described as rabid anarchists, and many of whom were respectable professionals, had anti-fascism stickers on their walls and windows.

'So how does Huxley use smell in *Brave New World*?' I ventured, wary of wasting our short time together talking dogs, without any idea of where the museum was and fearful we'd arrive there before I could find out about his thesis.

'He employs it as a technocratic means of social control. Pretty amazing, when you think about it, the way he anticipated a development in the olfactory realm that has only recently been discussed with any degree of seriousness.'

'I've not really heard it talked about outside specialist circles.'

'Oh, a number of people have called the twenty-first century *'das Jahrhundert des geruch'*, the century of odour. The nose has gained in stature. This gain is reflected, for instance, in a more aggressive stance by the perfume industry and the exploding artificial fragrance and flavour business. I think the first newspaper reference I saw to the social implications of environmental fragrancing systems was in the 1990s, in the *Wall Street Journal*. In it, executives of a leading fragrance manufacturer acknowledged that smells of the future could be potent enough to put smell suppliers in competition with the pharmaceutical industry. A former chairman acknowledged that the dark side of these developments was mind control. And a number of companies have found that adults are fond of the scents in products they used as a child, like baby powder, and are looking into fragrancing new baby products with this in mind. The fact that early childhood exposure to certain odours forms attachments that go on forever is already built into their profit calculations.'

'So are many people using these environmental fragrancing systems?' I asked, distracted from *Brave New World* by his insight into the world around me.

'Well I haven't really kept up to date with developments. But already in the 1990s, these systems were operating in hotel resorts, healthcare facilities, retail stores and office buildings in the US, Europe and Japan. The concept of adding scent to the architectural design process was just beginning to develop then.' I thought about the counter-terror conference and the talk about incorporating security features into our physical environment. The ventilation systems were already in place.

'And Huxley predicted these developments?'

'Very much so. You know how odours came to be associated with disease in the seventeenth, eighteenth and early nineteenth centuries and the authorities sought to eliminate them?'

'Yes.' I'd seen pictures of the plague doctor's dress that had haunted me in my sleep. The head and body fully cloaked, only one of the doctor's eyes was ever visible through the bird beaked facemask used to fend off the odours of the sick. The enormous pelican-like beak would be filled with protective aromatics. So paranoid were some physicians about contracting illness from the volatiles of disease that, according to a source cited in Alain Corbin's *The Foul and The Fragrant*, one doctor claimed to notice, after having spent a day in the company of corpses, that his 'wind produced an odour very similar to theirs'. In fact, the discovery that black absorbs odours more readily than white led scientists to recommend that physicians change the colour of their black cloaks and that the walls of medical establishments be painted white.

'Well,' Hans continued, 'in the *Brave New World* smells are no longer something to be eliminated by the system, but rather something to be exploited by it. They become a tool for manipulation of the masses.'

'How?'

'The old world stinks, the new is full of pleasant scents, and the olfactory is used as a political instrument of control over the individual. In the new world, olfaction has developed its own technological apparatus and can be implemented effectively for political and psychological purposes.'

'What technological apparatuses?'

'The perfumed plumbing system, the soma vapour guns, the scent organs. Smells are a communal turn-on, directed at nobody in particular. In *Brave New World* we see the mass application of de-individualised olfactory standards in the name of maintaining social stability. The brave new world is thoroughly sanitised and standardised, medically and hygienically controlled. The public and private spheres have been cleansed of unpleasant odours and are periodically infused with good smells from specialised equipment.'

'It's a bit like the smoking ban and pub chains,' I said. I'd argued endlessly over the implications of the smoking ban with friends and pseudo-libertarians. What got to me most was the lack of resistance to it from smokers. 'It'll help me give up,' they'd say. But in opting for external control, weren't they giving up the ghost? I'd throw in the fact that Hitler was the first politician to propose a smoking ban for rhetorical purposes, as Werner Pieper had told me he was, but my real fear was that it was the sign of a trend that would see us obliged to jog in the morning on the government's orders. We'd need to take care of ourselves for the good of the *Volk*. I'd heard talk about how midwives are advising parents that if either smokes, the baby should sleep in a separate bed because during the night, nicotine will 'escape from their blood' and 'penetrate the baby's skin'. Even the smoker's breath was described as 'potentially harmful' to the newborn. Things were definitely getting out of hand. Bill Durodié, senior lecturer in Risk and Corporate Security at Cranfield University argues that we are beginning to see everything through 'the prism of risk' and that our growing intolerance is a symptom of a 'mass psychogenic illness'. Whatever the causes, the result was that if I decided to open up a smoke-filled jazz bar, I *wasn't allowed to*. Most people agree that any risk from passive smoking is too small to panic about. The principal objection to smoking in public among those I've asked is 'the smell'. In fact, to everyone's surprise, the stench from pubs has become unbearable as a result of the smoking ban and they are now being pumped full of synthetic scents, including some sold as subliminal mood enhancers to encourage beer consumption.

'Yes.' Hans laughed. 'As Mustapha Mond says in *Brave New World*, "There isn't any need for a civilized man to bear anything that's seriously unpleasant".'

'Why was Huxley so preoccupied with the growth of science and technology and the impact it would have on autonomy and liberty?' I asked him.

'Well, if you think about the time he was writing *Brave New World*. It was between the wars. The catastrophe of the First World War can be seen as the culmination of a belief in technological feasibility that, combined with a reckless imperialism, spells utter disregard for the individual human being. For the generation of the first two decades of the twentieth century the experience of the war with its mass deaths – due, precisely, to new technologies such as the machine gun, the aeroplane, or poison gas – was a formative psychological experience.'

'I guess so. Huxley became very concerned, didn't he, about military involvement in scientific affairs and the need for robust ethics in the scientific community?'

'A lot of writers have been. The fear is that in subjecting the human to microbiological analysis, the individual becomes a mere atom, and atoms can be split. Science divides up the individual on the operation table, saving some of its functions, discarding others.'

We walked in comfortable silence for a few moments, as Hans took note of the names of the roads we were passing.

'Of course,' Hans said, 'Huxley wasn't the first to note the totalitarian potential of odours. As early as 1911 the German writer and philosopher Salomon Friedlander wrote a sci-fi piece about purifying the air of the whole world. It's a totalitarian project and large parts of the world get exterminated in the process.'

'Why the correlation between odour and totalitarianism, do you think?'

'Fascist politics is, among other things, politics of the body. We have the rhetoric of health, contagion, and disease of the *Volk*. The medicalisation of genocidal practices, and at the furthest end of the spectrum, the concentration and extermination camps that bear down directly on the bodies of the prisoners. The olfactory medium serves as an excellent entry point into the politics of the body.'

'So, it's because of the power of odour that totalitarianism and olfactory measures go hand in hand?'

'Sure. I mean, the body as the locus of sensory perception can be viewed as the provider of all that we can be sure about. Even Hegel,' – Hans indicated we were turning left into Ruschestrasse – 'compared the communication of pure knowledge to the dispersion of a scent in a non-resisting atmosphere. It's an all-out contagion. Therefore there is no defence against it.'

Hans stopped and looked at his street guide. 'It should be round the next block.'

'Great.'

'But of course, it was Patrick Süskind's *Perfume* that really put olfactory perception on the popular cultural map of our time.'

'The story of a murderer.'

'Yes. It's funny, people never really think of *Perfume* as being about anything other than smell. In fact it's a novel of fascination, even obsession, similar to the fascist obsession with charismatic power.'

'Do you think maybe Süskind, in his novel, sought to portray smell as the human soul?'

'Why do you say that?' Hans turned towards me and raised his eyebrows expectantly.

'Well, Grenouille has no personal odour of his own and as a result no empathy from other humans; they don't even notice when he's in their presence. It's only when he starts adorning himself with the scents of other people that he is acknowledged by society. And the women he tracks down with his nose die when he extracts their essence for his perfume collection. And in the end, when he unstoppers a bottle containing the scent of however many young virgins he killed, amongst the multitude who have gathered to witness his execution for their murders, the crowd is overcome with awe and worship him as a religious apparition. Like some sort of Holy Ghost.'

Hans nodded encouragingly and turned his head away from me momentarily to press the button on the pedestrian crossing.

'Could be,' he said, his pointed shoes elegantly traversing the zebra crossing.

'And,' I said skipping across the road after him, 'I came across this book a while ago, by a German guy called Gustav Jaeger. It's called the *Origins of the Soul* and apparently his thesis is that odour is the soul of a being. Anyway, I wasn't going to bother following it up because it's written in nineteenth-century German. But . . .'

As I told him about an article Werner Pieper had sent me, reporting on Jaeger's great-granddaughter's objection to similarities between Süskind's novel and *Origins of the Soul*, we turned off the main road into what looked like an English council estate.

'How interesting!' he said. 'I've not heard of it. Have you read it?'

'No,' I said, following him up the marbled steps into a heavily graffitied concrete block. 'It's in German. I'm hoping a friend here will translate it for me. But I did come across a patent taken out by Jaeger. It describes the preparation by "neutral analysis" of a scent-extract from the hair of healthy women. The hair is cut into small pieces and ground, first with lactose, then with water and finally with alcohol. It says homeopathic dilutions of the extracts are energising and animating.'

'Interesting. *Origins of the Soul* you say? I'll get a copy and look into it.'

A small statue of Felix Edmundovich Dzerzhinksy, who, Hans explained, was the founder of the Bolshevik secret police force, greeted us on entry to the museum.

Hans leaned in to the receptionist's window and smiled broadly.

I stared at a portrait of Lenin someone had spray painted over with the words 'The sleep of reason breeds monsters'. The scruffy picture looked strangely out of place in the otherwise sparse environment of this shiny black tiled room.

It didn't take long for Hans to charm one of the ladies who worked in the museum into giving us a guided tour. Unfortunately, it was in German, and as hard as I looked into the intensely green whirlpools of the snake-like eyes of the guide garbling at me, I couldn't understand a word of the everlasting grief she felt from living under the Stasi regime.

But I took in the exhibits. Cameras and listening devices hidden inside bird-houses and wooden logs, car doors, the insides of which were embedded with infra-red cameras, hung off the walls haphazardly. A whole room was dedicated to the Stasi dogs and their 'political operative work'. The dogs, the information cards explained, were used to track down anarchists and political opponents. 'Odour clues' would be sampled from political fliers, letters and pamphlets by holding a cloth on the object in question for a minimum of 30 minutes (unless the crime had been committed over 24 hours ago, in which case it should be held there for considerably longer). The sample would then be compared to second odour clues, surreptitiously obtained from absorbent pads secreted under the chair on which the suspect was placed during a fearful interrogation ordeal. The scent jars were lined up inside a glass exhibit box along with the forensic tools of the odour clue collectors. The accurate sampling of the odours was said to require special equipment; a dust cloth to be impregnated with the scent, sterile tweezers and aluminium foil. Even in the GDR, scent evidence from a dog was not officially condoned as evidence sufficient of itself to convict a suspect, though hundreds were convicted on the basis of it. Each jar bore the name and crime of the soul within, the deodorant with which the body had been adorned, its activity within the last 24 hours, and the names of any animals with which it had been in contact. The final exhibit panel bore as its title the overriding objective of the Stasi operation: the transparent citizen.

Hans and I left the exhibit rooms and walked out into the hallway where huge black and white photographs depicted the storming of the Stasi offices in Leipzig on 9 October 1989. Seventy thousand rapturous faces, illuminated by candlelight, were calling for an end to the Stasi spy network. After several years of living in fear as a result of being spied upon, they suddenly decided that enough was enough and confronted the Stasi *en masse*. Overwhelmed at the sudden lack of fear among their subjects, the officials inside the building began shredding their files in a desperate panic lest the extent of their surveillance activities be publicly unveiled. Piles of papier maché littered the corners of the museum hallway, relics of the shredded Stasi files.

'So,' I asked Hans, when the time came to say good-bye, 'do you think I should be afraid of environmental fragrancing systems?'

'Do I think you should be afraid?' he asked in surprise.

'No. I think it's an interesting area to look at because essentially, the investigation of the olfactory is the investigation of everything else. But even the most recent physicochemical research is at a loss to explain why certain molecules of a very similar structure produce sensory impressions of a completely diverse nature. Even in *Brave New World*, in which olfaction is used as a control mechanism, olfaction is also what results in its downfall. The political manipulators of *Brave New World* control and manufacture the good part of the olfactory spectrum; the bad part, however, emerges against their will, escaping, as it does, from acts and processes of an ethically questionable nature.'

I thanked him for his time and we parted ways. Hans was going to meet friends for dinner. I was getting the train to Mannheim for the next day's 'Non Lethal' Weapons European Symposium.

As the afternoon set in and I boarded my train, it began to drizzle. The atmosphere shed its tears across the window as I sailed out of the mist towards the south-west.

Kids undergo sniff test

Air Sense,
12 November 2007

Minneapolis' Children's Hospital has been studying how kids of different ethnic backgrounds interpret different smells. This study is among the first scholarly research on the effects of ethnicity and gender on kids' attitudes toward fragrances. The study involved kids from Hmong, African, and African-American backgrounds, sniffing fragrances that are being used in medicine to help deal with anxiety, nausea and headaches.

CHAPTER TWENTY SIX

Dragonfly out in the Sun

*The fool does not lead a revolt against the law; he lures us into
a region of the spirit where . . . the writ does not run.*
  Enid Welsford, *The Fool*

Two days later and Käte's bedroom was awash with
sunshine when I woke to the sound of a church bell
chiming. I congratulated myself on having taken my leave
from the less-than-lethal weapons symposium. At the close
of the first session I'd feigned a bout of food poisoning
and gotten away with the conference goody bag. As I
wouldn't be attending the rest of the symposium, I was
only required to pay the cancellation fee. I'd decided I had
all I needed of the proceedings in the book they'd given me
on arrival. If I felt like reading up on the neural impact of
high power microwaves, the psychological impact of multi-
sensory assaults and psychomotor sedations, or the direct
electromagnetic stimulations of pain receptors, I could.

And if the prose was too dense there were black and white photos illustrating the electric stimulation of cardiac tissue, gun-shot guts picturing the 'viability' of the large intestine, and a shackled pig being injected with a substance that resulted in destruction of the 'biological object', all available for me to examine at my leisure. I even had graphs, outlining the impact of 'voltage impulses acting on the bio-objects' and the 'higher levels' of their nervous system (spinal cord and brain apparently), and strange computer-modelled analyses of crowd psychology.

Käte had picked me up at Mannheim station two nights before. She'd been disappointed to hear that my visit would be spent at a less-than-lethal weapons conference. I'd been growing increasingly irritated by this fact myself. I enjoyed her company and hadn't seen her for a considerable time. Germanic in her logic and obsession with order, her Arabic heart delighted in sensory pleasures, her pursuit of which was almost child-like. She depended on her freelance work as an English teacher to business executives for a regular income but she was always working on other random projects that came her way; book translations, pop band interviews for television programmes, sales of luxury goods (from BMWs and pharmaceutical drugs to €2000 glass-blown bongs) and personal assistance to visiting celebrities. Her English was perfect, as was her German, and her sitting room was filled with novels, English on the left and German on the right, all shelved in alphabetical order.

The less-than-lethal conference was being held in Ettlingen, a small village some two hours drive away from Mannheim.

'So,' she said, after telling me how to get the train to Ettlingen in the morning and passing me her tobacco pouch, 'first we're going to Werner's to pick up this book he's found for you?'

'Yes, but I'm afraid we can't stay long. I'll have to get up very early tomorrow.'

'OK,' she said in a tone that suggested I was being unreasonable.

'It's stupid,' I'd agreed as Käte leaned over the steering wheel to get a closer look at the motorway signs. 'I met this anti-arms campaigner last month and I told him about this smell conference I went to and about how weird it was to see guys from the Ministry of Defence chatting earnestly with perfumers. He was really interested to hear this because he'd been out to the Monell Chemical Senses Center in Philadelphia where they are designing malodorants for use as less-than-lethal weapons. I told him I'd like to find out more about less-than-lethal weapons. So he offered to introduce me to one of the "Men in Suits".'

'The Men in Suits?' Käte had asked in a disparaging tone, wrinkling her pixie-like features into an almost angry expression and taking the cigarette I'd rolled her out of my hand.

'I know, it sounds silly.'

Käte's hazel-brown eyes remained fixed in deep concentration on the road ahead, her thick black eyebrows furrowing deeper into her face as I'd continued.

'But apparently they're legendary in the UK for their ability to get into high level security conferences. They have a fictitious security company. They go in, get all the information and give it to activists like the guy I met.'

'But Amber, this stuff is available on the internet. They aren't secret these things. I teach English to a guy who runs a security firm. It's all on the net. I've got his Power-Points at home.'

'I know. I know. That's why I feel stupid forking out €1200 to go to this conference. Particularly if one of the Men in Suits is there anyway.'

'€1200?' Her face dropped. 'That's silly Amber. Think what else you could do with the money.'

'Ahha. I know,' I said, remembering how little of it I had and admiring the countryside we were driving through. 'But one of the Men in Suits told me his partner was going to it and that I'd never get in because of the security. He said he'd fill me in if anything came up on malodorants.' I'd paused, ashamed of how I'd got myself into this situation and beginning to regret it. 'He even told me there wasn't anything about malodorants on the programme, and there isn't. I was just so irritated,' I'd confessed to Käte, 'that he didn't think I'd get in, that I applied for a place. Anyway, now I'm obliged to go or pay a €300 cancellation fee.'

Käte shrugged silently.

'I know a lot of this stuff is on the internet,' I said, trying to justify my decision-making process, 'but it can be interesting to talk to the participants and listen to what they have to say.'

'Sure,' she'd shrugged. 'It's just a shame with this lovely weather that we're having you'll have to be inside, listening to different ways of inflicting pain.' How accurately she'd predicted the content of conference talks, I mused, stretching out in her empty bed, listening to her singing in the kitchen. 'Pain is the most common effect. Its purpose can be described in various ways – to incapacitate, to distract, repel.'

'Actually,' I'd said to Käte in the car, 'I hoped that maybe you could join me for some evening drinks with these men. They're mainly police officers and military types. A lot of rich arms dealers.' I knew she'd had a lot of experience with these sorts of people from teaching military officers on the United Arab Emirates–Germany exchange programmes. And most men were immediately attracted to Käte. She was kind and playful in their company and sexual confidence emanated heavily from her body's curves and movements. She could have got them to spill the beans on malodorants.

'Well you know Amber, these military guys are still human. They are no different to you or me really.'

Maybe she was right, I thought, as I listened to the church bell and wondered what Käte was up to and whether there was any coffee in the house. But it hadn't felt that way at the symposium, under the strip-lights in the confines of the stuffy basement of the chemical institute. The atmosphere was so stifled I'd found it difficult to breathe. I'd sat there on the middle row, behind an assortment of soft drinks I was expected to share with the unsmiling man next to me. He'd already polished off the only bottle of Coke. I fiddled with a bottle of mineral water territorially, listening to a German scientist outline the impact of a weapon for use in the Global War on Terror (abbreviated to GWOT in the proceedings): 'It will produce a biomedical response, e.g. large organs moving about in the body which will trigger an injury mechanism.' I began to feel nauseous. Perhaps some hungry creature's volatiles had wandered into my amygdala, catching me unawares.

Käte had pointed out of the windscreen, drawing my attention to the extraordinary redness of the sun setting between the hills. I'd eyed the setting sun with suspicion, wary of the symposium and the people it would be attracting. Perhaps, I thought, Mother Nature was already playing the security game and the area had been put on red alert for the less-than-lethal event.

'This area is famous for it. It's because of all the pharmaceutical companies we have,' she said.

'I think this is it,' Käte had said, swerving out of the way of a pretty train through the windows of which I saw smiling couples drinking glasses of champagne.

'It's the party train,' Käte had explained. 'It only

operates on Sundays. You can take it all the way from Heidelberg to Mannheim.'

Käte had turned right and driven through a small town. We'd driven up a steep hill, past a small medieval church and a waterfall, and pulled up outside a vine-covered two-storey stone house. Käte switched off the ignition and rolled a cigarette.

'So where exactly are we?' I'd asked, not having been to Werner's house before.

'Birkenau. Of course Werner doesn't call it Birkenau because another Birkenau was part of the name of the concentration camp Birkenau-Auschwitz. He calls it by the old name of Lörbach, so it took me ages to find it the first time I came here.'

We'd got out of the car and walked up the crumbling steps to a wooden door on the upper storey. The wall beside me was alive with bees buzzing manically amongst the flowers.

Werner opened the door to a yellow painted room, smiled broadly and outstretched his wool covered arms.

'Good to see you.' He'd hugged us both.

'Where are all the bees from?' I asked.

'I don't know.' His long grey dreadlocks flapped round his shoulders as he shrugged. 'I don't follow them.'

His house was as inviting and beautiful as I'd expected it to be from the colourfully adorned writing paper on which he'd written to me. He'd led us through the freshly prepared food-filled kitchen and into his living room. We waited on a red sofa in the candlelight while he fetched us apple juice, made by his farmer-neighbour. He placed an earthenware jug and three cups on the table with some cake, and sat himself on a large fisherman's net that swung from the ceiling like a hammock.

'So,' he said. 'I got your book.'

'And?' I asked excitedly. Werner had been collecting

and publishing bizarre books on all manner of eclectic subjects for over thirty years and I'd hoped he'd be interested in *Origins of the Soul,* which he'd not heard of until I'd mentioned it. I was also hoping he'd have read it and be able to tell me about it in detail.

'It's weird. I don't like to look at it.'

'Oh?'

'It was written in Imperialist times. The language he uses, it is . . . I don't like it.' His facial expression suggested he'd detected something disturbingly dark and evil in the book's passages and the way he flinched before reaching to get it from the shelf beside him made me think he didn't like to touch it.

'He talks about Jews having a certain smell. And other races. You can see the thought processes that led to the Third Reich creeping through it. So,' he'd handed it to me and slapped his thighs, 'I leave it to you.'

It had taken us longer than scheduled to leave the time capsule that was Werner's house. He'd regaled us with stories for several hours; how he'd hated Heidelberg when he first moved there, 'then I ended up dropping acid with a load of GIs and swimming in the Neckar river with them and I realised that there was something special about the place.' Werner saw himself as a psychedelic warrior, fighting mind wars on behalf of the 1960s hippy generation.

He'd started dealing acid to the GIs. 'Buyers would normally want to test out what they were buying so this one night I was sat on the old bridge with a GI and we took a tab each. Just as it was kicking in I pointed to this round object that could be made out next to the castle behind the trees on a hill. I told him the townspeople would go there and smoke from a giant chillum at 1 a.m. on the dot. All of us dealers used to sell them hashish in old wooden pots and bottles – like it was an old German tradition. In fact the only reason so many dealers were in Heidelberg was because the US military headquarters are based there. So the psychedelic movement in parts of Germany was indirectly sponsored by the US army. But that's another story.' He'd waved his arms in the air and smiled. 'And then he looks and he sees a huge puff of smoke emerge from behind the trees. It was the steam train – I had spotted it entering the tunnel and so knew what time it would be passing the huge chimney on that hill, actually called the King's Chair. Well – it blew his mind. He loved it.'

'Well,' Käte had interjected. 'I wonder about the sheep up in those mountains. Last autumn I thought I saw them eating magic mushrooms and I wondered if this is what they were really doing and the next day I was sure that this was what they had done.'

'Why,' I'd asked. 'What were they doing?'

'Oh. I don't know. Just standing there. Looking . . .' pausing in search of the English word 'sheepish', she clicked her fingers and burst into laughter.

'I have a book on the drug habits of animals,' said Werner. There weren't many topics he didn't have books on.

Werner asked me for the reasons behind my desire for Jaeger's book, which I tried to elucidate but got sidetracked into discussing the increasing security measures being implemented in the UK.

'Ah,' he said. 'That reminds me. I visited England recently. As you know, I always travel by train.'

In his hash-dealing days, Werner, who never got a driving licence, preferred travelling by train, sitting in the buffet-wagon, sprinkling both wagon entrances with pepper. He claimed that 'no dog sniffs anything else for some time after hitting pepper'. His car-driving friends,

however, 'preferred to keep a ham sandwich on the dash-
board of their cars'. During his time as a dealer, he acquired
a passion for travelling by train, enjoying the sensation it
gave him of getting to 'places by foot'.

'I was just about to cross passport control at Waterloo
when a customs person comes up to me. "Excuse me Sir,"
he says. "Are you a practising Rastafarian?" I look at him
and I say, "I will not answer that question. I do not discuss
matters of religion with people like you."

'"What do you mean, people like me?" I looked at him
and I said, "I mean people who wake up suspicious of
everyone and everything."

'"Do you smoke marijuana joints?" he asked. Well, I
have always smoked hashish. I have never smoked grass,
so I could tell him from my heart, "I have never smoked
marijuana in my life." Like Dylan says, "To live outside
the law, you've got to be honest."' He winked and laughed
before an expression of seriousness resumed its place on
his wizardly face.

'He was not happy with this answer. So now I start
to take my clothes off you know? "We can do this in the
open." Well, this really unnerves him and he tells me
"Stop. Stop. It's fine. You can go." Ahh – a funny thing
– another story here – only once that I remember, have I
been inside a lunatic asylum.'

I'd had to stop him, wary of the hour and the reality
awaiting me. We said our goodbyes and Werner opened
the door. Käte cursed the weather. It was pouring with
rain outside. Germany, like England, was having its sunny
spells interrupted with torrential downpours this spring.
Werner put his hand on her shoulder. 'I don't complain
about the weather any more,' he'd stopped her and said.
'Since I was a little boy. One day, after it had been raining
for five days I complained to my grandmother, "I'm bored!"
She tells me, "You should learn to accept what the weather

brings" and she wagged her finger at me. "And be glad that we cannot change it".'

Käte wasn't convinced. It was a long drive home. We hugged and kissed Werner goodbye.

'Oh,' he'd said as we hovered reluctantly in the doorway. 'I'm sure Käte will tell you this but anyway. You do know, don't you, that Jaeger means "hunter"?' He'd hissed the word melodramatically and laughed, waving goodbye as we descended the steps.

Back in Mannheim, I'd gone straight to sleep but Käte had stayed up for hours, galloping red-eyed through Jaeger's *Origins of the Soul*, high as a bat on skunk, and typing furiously.

The following morning, when I was due to get the train to the symposium, I'd discovered that my suit was moth-eaten. Käte had clambered sleepy-eyed into her wardrobe and given me one of hers. 'Good luck,' she'd croaked, 'see you tonight. Call me if you want me to pick you up. I finish work at seven.'

In fact, I'd arrived back before then and we had dinner in a sushi restaurant. I'd fallen straight to sleep when we got back, tired from my day trip to Ettlingen.

'Good morning!' Käte came into her room, dressed in a long and loose-fitting short-sleeved red dress. She opened her wardrobe and threw me a similar version in orange. 'You up for some breakfast and a trip to the lake?'

Over breakfast, doubts began to resurface in my mind over whether I should have remained at the conference. I was glad to have spared myself the €900 and gotten away with the proceedings and was fairly confident that I'd witnessed the most interesting part; the discussion forum that kicked it off. But the doubts niggled away at me. There'd been so much I hadn't understood. Why did the scientists find it difficult to know what the military wanted? Why was the specification of desired effects compared to the opening of a Pandora's Box? I remember the various audience answers to the chairman's question, 'Is a magic bullet possible?'

'Yes,' an elderly American man had said, 'we see it with hunting animals – it is possible to make a human unconscious for half an hour.' 'First,' piped in a neuroscientist, 'we need to specify the behavioural response desired.' 'A magic bullet is possible,' chimed a small Chinese man. 'It is a question of price and time.' He'd reminded me of the shop merchant in the film *Gremlins*. 'We have the components; we just have to mix them up. Takes time and money.'

'Then you wouldn't be having this lovely breakfast now,' Käte rebuked me. She was right. This was the best breakfast I'd ever had. Smoked salmon, mango, avocado and toast with fresh orange juice and an egg.

'Whether it be giving someone the evil eye, or a laser beacon, it is about escalating the will of the person using the device,' another member had explained, leading the neuroscientist to conclude this particular topic with a sentiment I failed to comprehend: 'I think you have found your magic bullet: it is knowledge space and not a device.' What did he mean?

'Now come on. Let's go. The sun's out and there still won't be many people at the lake at this time.' I think she could tell that my nerves were frayed with all the talk I'd heard of frying them.

Perhaps it was better not to know what the neuroscientist had meant. Downing the remains of my orange juice I recalled the last thing I'd heard before I left. 'As soon as they learn how the devices work – we have to escalate our machinery'.

Out on the road, Käte pulled me into a pharmacy and purchased some moth pheromone traps before getting in the car. We drove through Mannheim, Käte enthusiastically pointing out of the windows informatively. 'That's the industrial sector. Most pharmaceutical companies have their headquarters there,' she said, pointing below us as we took the corner of the *autostrasse* into the centre. 'Even up here we have a large number of factories. My friend Dieter lives over there,' she pointed beyond the town centre. 'People don't live there for long because of the smells from the chocolate factory. The streets change their smell every day depending on what flavour confectionery is being made.' We drove along the main road towards the Neckar river past a line of tall concrete buildings. 'Those are US army barracks.'

'They're huge!' I commented.

'Huge,' she concurred. 'They have their own bingo halls, restaurants, bakeries. The currency is US dollars.'

'What are they doing there?' I asked.

'Occupying Germany.'

I laughed.

'That is all they're doing.' She frowned.

Half an hour later, Käte turned left into an asparagus field. Pickers lined the dirt track we drove across. As we approached the woods, she turned right into a small clearing for cars, got out and grabbed a couple of rolled up mats and a notebook from the boot.

'So,' she said proudly, leading me through the grass park and past a terraced bungalow serving food and drinks. Empty hammocks overlooking a huge lake lined the wooden support structure of the terrace. Taking me by the hand and leading me into the woods, she said, 'This is Die Schilicht.'

I followed her through the overgrowth until we reached a secluded sandy cove, at the foot of the beginnings of the lake, where she dropped her book to the ground and laid out the mats. Käte laid the mats down on the dry mud of the bank and I lay under the shade of the trees. The shallow green water stretched out into unimaginable depths beyond our feet. Dragonflies jumped from reed to reed, fiery green in the sun.

'You've got to be careful of ticks,' she said.

'Really?' I asked, jolted out of my dazed appreciation of the scenery.

'Yes. They live in the grass and carry viral diseases that attack your nervous system and give you, ah *shitza*, brain disease um . . . meningitis! I think that is the English word. Yes. Meningitis.'

I must have looked horrified as she quickly assured me that there were unlikely to be any in this particular spot.

'Go for a swim,' she said.

'Are there any ticks in the water?'

'No!' She batted her hand at me and winced. 'Although there is a myth that a giant, ancient pike lives in the deeper parts. But it's just a saying.'

Käte had a hard job persuading me to get in the water after such warnings, but eventually I did, walking delicately over the twigs until the ground disappeared below me and the lake broadened out between huge mountains barely visible from where we had been sitting. I ducked my head under the ice-cold water, re-emerging from the depths in delight as I soaked in the view.

I swam back to Käte, and lay down on the mat, enjoying the heat of the sun beating down on me.

A cuckoo flew low across the water, singing its song. A far preferable means to tell the hour by than the office clock, I mused as a smile tickled my warming cheeks.

'So,' said Käte, evidently pleased to see me happy and relaxed but eager to please me still more. 'I finished my translation of the passages you said you were interested in from Jaeger. Now that you are dry and comfortable,' she blinked and bowed her head, handing me her typed pages.

'Thank you,' I said, amazed by her generosity, and

started to read. 'I believe that I have solved the problem of the animal soul' it began.

> It is evident that the soul itself must be a specific substance . . . there is only one such group . . . those substances which contain the specific odorous and savourous matters of an animal, because these substances alone are of a completely specific nature . . .
>
> These specific substances are of the utmost importance for the functions of self-preservation and propagation. The food which every animal species selects depends on its specific smell and taste. No experience whatever is necessary in this respect, and the caterpillar, for example, creeping forth from its egg, recognises with unerring certainty the specific plants which form its food. When the kittens of a cat are shown the image of a dog it makes no impression whatever upon them; but if a living dog be rubbed with the hand, and the hand be brought afterwards into contact with the nose of a kitten, the effect is striking, because the cat smells and instinctively recognises its enemy. Hence it is plain that the animals mentioned are in these cases only guided by their chemical senses. Odour is the means by which we know what food or what mate is best suited to us. Our individual odour must be in harmony with the food we eat and the partner we choose.
>
> Smell is the essence of the substance which is the soul of living beings. We all have our own individual odours. The more mature and developed the individual, the more we can tell from their odour. In youth, all that can be told by the odour is the racial origin. The taste of a new-born animal, for example, is not very tasty at all. The older the animal, the

stronger the taste. Equally, Jewish children do not smell as strongly as the adults. All members of the same race or species will have a familiar odour to each other.

This must have been the passage that disturbed Werner. I had to admit that there was something creepy in his use of the word 'tasty' in the observations on newborn animals that immediately preceded his racist remarks. Its absence in his discussion of Jews left a bitter taste in the mouth, as if he had omitted the most unsavoury details of his experiments. But I suspected that Käte was right, and that far from being a cannibal, Jaeger simply employed arcane language to explain his scientific and philosophical observations.

What does not taste or smell will not affect our essence. Human behaviour is not determined by nerves, but by smell molecules. If my proposition that the soul is a specific chemical substance is correct, then it must be subject to the rules of metabolism. The material to build it will be taken from the outside.

In other words, Jaeger was arguing that our soul could be influenced by external odours, alien souls and, it occurred to me, malodorants.

The emotions, the soul of the brain, all have different smells. You can smell and taste fear. Fear reeks. Whether we feel fear will depend on the harmony between our soul matter and the external stimuli. The mind is consciousness and the soul is the non-conscious self which is used for self-survival and propagation. The observation that when an animal is frightened – especially when we have to do with mortal fright and terror – malodorous matter emanates from its body, is an old one. Dr Albert Günther, director of the zoology department of the natural history museum, sought to research 'todesangst', the fear of death, by killing a cat in the British Museum. The stench of the cat's fear, urinous in character, was difficult to get rid of. The fear of death was what caused the stench.

I tried to imagine this eminent zoologist, alone at night in the grandeur of the British Museum, hacking away at a cat, noting the various odours emanating from the terrified creature at different stages of decay in its biological system.

Fear matter in the body and its role can best explain the practical application of my theory of the soul. There is a specific smell to fear. The origin of the smell is in the brain of the organism – if the brain of a suddenly killed animal that had not time to waste its fear matter – if you take its brain and smash it up in a bowl and add some acid – then you get the smell of fear. The smell of fear comes from intense nerve stimulation of a disharmonic kind . . .

It has a direct effect on our bodily juices and travels through the body. Its effect is not localised but general – on the muscles, on our breathing, on our heart, in the intestines – even our hairs can be bleached white by this substance. The body seeks to expel this substance and psychic problems result if it is unable to escape . . .

All feelings are the result of the interaction between the soul substances and the external substances. So, if there is a boar in the room, its odour will interact with your soul matter and, as a result, you will feel the fear. The individual's soul

matter, their self-smell, is the will, which is the head
of the machine which is the body. One's character is
determined by the odorous matter of the soul. The
nature of our soul and all psychic reactions are based
on the metabolism of odours . . . In the house of the
body, smells play the role of headmaster. Instinct and
drive are the phenomena of the soul.

The ghost in the machine was a smell. It decided who we
were and what we did. There was somebody watching. It
was the master. But who or what was it?

Character, which is contained in the smell molecules of the soul, is passed from one generation
to the next. Dog breeders, who are ridiculed by
physiologists for their claim that they can successfully breed dogs to exhibit certain character types,
are in fact correct. Character is inherited, from smell
matter. If further proof of my theory is needed – the
etymology of words for soul are all smell related.
Moses blew life into man, and we talk of creative
inspiration. The Ancients knew that smell was the
soul. Alcohols, because of their strong smell and
angry volatiles, are called 'spirits' . . . The individuality of our breath is the route to our soul.

So, it was pretty clear these olfactory chemists were
after my soul. First they want to extract and examine my
essence. Then they want to mix it with their vile sulphurous
compounds reducing me to a fearful, compliant version
of my former self. I could picture row upon row of scent
jars in the police laboratories.

It is for this reason that clothes worn are of psychic
importance and woollen clothing is advisable.

US accused of making insect spy robots

Daily Telegraph
Tom Leonard
10 October 2007

The US government has been accused of secretly developing robotic insect spies amid reports of bizarre flying objects hovering in the air above anti-war protests.

. . .Vanessa Alarcon, a university student who was working at an anti-war rally in the American capital last month, told the Washington Post: 'I heard someone say, "Oh my God, look at those." I look up and I'm like, "What the hell is that?" They looked like dragonflies or little helicopters. But, I mean, those are not insects.'

Bernard Crane, a lawyer who was at the same event, said he had 'never seen anything like it in my life'. He added: 'They were large for dragonflies. I thought, "Is that mechanical or is that alive?"' . . .

The CIA secretly developed a petrol-powered dragonfly drone back in the 1970s but the 'insecto-hopter' was considered a failure as it couldn't handle cross-winds. The CIA refused to discuss its subsequent work but it is known that the Defense Department has been funding research into inserting computer chips into moth pupae to create 'cyborg moths' whose flight muscles can be controlled remotely. Although experts say there are still considerable technical hurdles – not least finding a way of protecting the creations from hungry birds – some concede it is possible that some agency has secretly managed to make something that works.

'America can be pretty sneaky,' said Tom Ehrhard, a retired Air Force colonel and expert on unmanned aerial craft.

PART III

CHAPTER TWENTY SEVEN
Fear and Freedom

In the office, the shatter-proof curtains – designed to ensure I didn't get sprayed with glass if the building got bombed – were (according to the stranger who had just entered the room unannounced) in need of replacement. As the man fiddled round the window, I found my thoughts drifting into novels that spoke about individual autonomy and security, and looking for scents in them. In Ken Kesey's *One Flew Over the Cuckoo's Nest*, Chief Bromden, a half-American-Indian, whom the authorities believe is deaf and dumb, tells the story of a mental institution ruled by Big Nurse Ratchet on behalf of the all-powerful Combine (a group of companies working together for commercial purposes). A major theme emerging from *Cuckoo's Nest* is the Combine's desire for ruthless efficiency and uniformity and its treatment of human beings as malleable dots that can be manipulated with technology.

All up the coast I could see the signs of what the Combine had accomplished since I was last through the country, things like, for example – a train stopping at a station and laying a string of full-grown men in mirrored suits and machined hats, laying them like a hatch of identical insects, half-life things coming pht-pht-pht out of the last car, then hooting its electric whistle and moving on down the spoiled land to deposit another hatch.

As well as explicit references to olfactory detection of fear amongst the inmates, there were more implicit references to odour weapons. Chief Bromden makes frequent references to environmental controls used by Big Nurse to maintain the 'clean, calculated arcade movement' of the patients. Kesey was clearly on to the concept of less-than-lethal weapons and much of the plot revolves around the use of electric-shock therapy to calm the patients and make them more compliant. The inmates discuss the therapy at length with McMurphy, a prisoner who thought he'd have an easier ride if he served his sentence in a psychiatric unit rather than a prison, and who is proved tragically wrong. Harding, a well-educated inmate belittled into fearful seclusion from his desires, explains how the Combine cottoned on to electric shock therapy.

After all, a man wasn't a cow. Who knows when the hammer might slip and break a nose? Even knock out a mouthful of teeth? Then where would they be, with the high cost of dental work? If they were going to knock a man in the head, they needed to use something surer and more accurate than a hammer; they finally settled on electricity.

Harding's explanation carries echoes of the European principle of proportionality that one shouldn't use a sledgehammer to crack a nut. The most prevalent coercive technology in the Big Chief's mind, however, is the fog machine. Purchased from the army by their hospital, it oozes mist in thick white clouds across the hospital floor, pulling him across it like a robot and nauseating the patients to the point that they can't stand up for themselves or exercise any cognitive function. This, it now seemed clear to me, was a malodorant.

McMurphy, who before excessive electric shock therapy plays the renegade role of seeking to derail the machine by unshackling his fellow inmates from the belief that they are enslaved, is initially amazed to discover that most of the inmates live in the institution of their own volition. He'd incorrectly assumed they'd been sectioned against their will. He fights desperately to enthuse his fellow inmates to take responsibility for their lives by pursuing their own dreams but only Big Chief succeeds in fleeing the cuckoo's nest.

> Nobody complains about all the fog. I know why, now: as bad as it is, you can slip back in it and feel safe. That's what McMurphy can't understand, us wanting to be safe. He keeps trying to drag us out of the fog, out in the open where we'd be easy to get at.

In J. G. Ballard's *SuperCannes* the protagonist Paul Sinclair accompanies his wife to Eden Olympia, a futuristic paradise for hard working high-flyers, where she has accepted a medical position. Hugely wealthy chief executives and academically brilliant scientists all live and work within its high-security perimeters. At first Paul is stifled by the environment with 'its moral thermostat set somewhere between duty and caution' until he discovers the emotional outlets employed by his neighbours. He discovers that the rich and powerful people behind the security 'crime-free' environmental design get their kicks from torturing and killing immigrants and low-life crooks outside the perimeters of the complex. Seeping into the synthetic atmosphere of Eden Olympia we read of 'pharmaceutical odours' emerging 'from a ventilation shaft'. As Paul eases into the way of life at Eden Olympia he loses sight of himself and caves in to the external and artificially imposed moral compass; 'the scent of disinfectant and air-conditioning suddenly seemed more real than the sweet tang of pine trees.'

I thought about life in the civil service, and the promise of a 'job for life' and decided that the time had arrived for me to resign, before it became too late. I informed management, who didn't seem surprised, even by my lack of notice, and packed my belongings.

Garth seemed disappointed, but said he understood. He said he'd seen it coming and he was pleased for me.

'If I'd been locked up in the prisoner of war camp in Colditz Castle, I'd have never made it out.'

I told him I'd miss him.

'You can visit me,' he said.

My bedroom now replete with books, papers and security company sponsored notepads, pens, backpacks and briefcases, I felt I was reaching the end of my olfactory journey. Ben Teckler's emails had tapered off in recent months on account of his increased workload. He didn't seem able to find the time to dig out the information held on odour conditioning of human behaviour at the MoD olfaction department. And a response to my FoI request told me that they didn't hold any.

Roger had disappeared. His email address and phone numbers were no longer recognised. Neither Doug the Buzz Man nor Bee Girl ever replied to my emails following up on the Swiss conference.

Geraldine, the horse-faced woman from the dog show, had kept in touch with me throughout the year, sending me (on loan) videos of police dog shows, newspaper clippings and magazines. She'd even taken me for lunch at the Kennel Club, a grand and stately building in the heart of London, round the corner from the Ritz. Its elegantly oak panelled walls portray pictures of the club's dog breed designs, dating back to its establishment in the middle of the nineteenth century. While Geraldine freshened up in the ladies', I'd wandered over to a table in the centre of the richly carpeted bar area to have a read of the newspapers. But they were all dog newspapers.

When Geraldine returned, we sipped on aperitifs as she teased me with her access to the inside story on dog handlers. 'They've told me stuff they can't tell their chief constables. I've got it all in my files at home.' 'What do I have to do to see them, Geraldine?' I'd teased back. 'I can't honey, I can't. If I say I'll keep something confidential, then I keep to that promise.' I believed her. 'And equally, if you were to show me some of your research, I would keep it confidential.'

After filling me in on the history of the Kennel Club, and its rocky relationship with the police dog handler fraternity – 'the club breeds dogs that look good, the police want good workers' – Geraldine treated me to an exquisite three-course meal in the strangely formal dining quarters of the club and took me to have a peek through the door of the boardroom. I glimpsed the long mahogany table, velvet coated chairs and large painted portraits of important-looking men with moustaches and dogs.

Truth be told, the UK dog handler community had given me a good summer. After returning from Germany, I'd met with as many handlers and visited as many police dog headquarters as I could in the free time provided by my working week. If they were Gustav's disciples, they did his bidding unwittingly, led by their silver hounds.

I'd almost been reassured about their use of mental health hospitals as training grounds. Apparently they were on good terms with the patients as a result of spending so much time there. In fact, the officers had made a lot of friends amongst the inmates.

'One guy,' a handler explained to me over the 1970s disco lyrics, 'Yes Sir, I can boogie', that blasted out of his police van's stereo as we cruised down the motorway with the occasional supplicating whine of the caged dogs asleep behind us, 'loved police equipment and we and the boys were always giving him bits and pieces – a hat or something, every time we visited. We ended up getting into trouble over that 'cos one day he went out into the community dressed up as a police officer! He'd managed to get the whole shebang over the course of our visits.'

I'd even acquired a suntan as a result of the time spent sitting on a car bonnet watching a British Transport Police training session in a repossessed car pound. They'd generously shared the contents of their picnic baskets with me as we watched novice flyers take off and land on a neighbouring airstrip. Occasionally the dog would pick up on the hidden substance, and I'd watch its telling tail wag enthusiastically over its behind in enunciation of its finds. They didn't know how or when their dogs were right. It was a complex business; wind currents, nasal health, motivation, setting, temperature, difficulties in communication, miscalculated rewards and outdated training substances.

Except for the strange habit of referring to themselves as 'Daddy' when talking to their dogs, I'd found their company pleasantly rewarding and had grown fond of the Dogmen.

So when I came across a reminder in the police and services dog club magazine for the UK Dog Seminar I felt I had to go. It was advertised as the biggest Detector Dog event in Britain, attracting law enforcement officials, scientists and government ministries from all over the world. It proclaimed the provision of a unique opportunity for police constable dog men to meet with hard-nosed scientists. It took place only once every two years in a remote Scottish castle. I emailed Geraldine to ask her if she was going. She told me she hadn't planned to but that as it happened she would be in that neck of the woods the very weekend it took place. She might well pop in and say hello.

Pentagon looks to shark spies for ocean research

ABC News Online
3 February 2006

The Pentagon is funding research into neural implants with the ultimate hope of turning sharks into 'stealth spies' capable of gliding undetected through the ocean.

British weekly *New Scientist* has published the research, which builds on experimental work to control animals by implanting tiny electrodes in their brain that are then stimulated to induce a behavioural response.

'The Pentagon hopes to exploit sharks' natural ability to glide quietly through the water, sense delicate electrical gradients and follow chemical trails,' says the report.

. . . The implants, controlled by a small radio transmitter, stimulate either the right or left side of a brain area dedicated to smell, causing the fish to flick around in that direction in response to the signal.

The next step will be to take this device outside the laboratory.

CHAPTER TWENTY EIGHT

The Castle of the Dogs

Dr Evil: You know, I have one simple request. And that is to have sharks with frickin' laser beams attached to their heads! Now evidently my cycloptic colleague informs me that that cannot be done. Ah, would you remind me what I pay you people for, honestly? Throw me a bone here! What do we have?
Number Two: Sea bass.
Dr Evil: [pause] Right.
    Austin Powers

I made the mistake of flying to Glasgow instead of taking the train, and didn't arrive at the castle until after dinner. It was still light outside when I made my way through the grounds, and up a small hill towards the reception area. As I approached the doors, a huge hairy Alsatian bounded out of a parked up police van and ran towards me. I struggled to keep my cool. Life in the secure world is all about hiding your fear.

Two feet away from me it stopped at the foot of a bush and sniffed the flowers; an appropriate response. A couple came up behind it.

'Beautiful,' said the man, turning back to look at the dog.

'Yes,' agreed the woman, a ponytail swinging behind her as we approached the reception doors, 'he is.' She spoke as the owner. 'Although he is also an extremely good attack dog.'

'Yes,' I interjected with relief at this opportunity to express what was evidently not paranoia. 'I'm sure he is. I . . .' I let out a small laugh, 'was a little concerned when I first saw him bounding towards me.'

The man frowned. The woman turned to look at me. 'My brother is like that.' She paused. 'Terrified they'll attack him and so when he sees them he runs. And then of course they chase him.'

'What was so natural about chasing after a person's fear in an attempt to pounce on them?' I asked myself. Cats and donkeys don't chase after you just because you're scared. Not even spiders do that.

We parted ways at the reception desk where I introduced myself to a lady behind the counter. As I selected my badge from those on display and dropped it into my bag, I turned to see Loretta walking out of a corridor and into reception. She had joined the MoD contingent at the Abode of the Devil a day after our arrival. She had never really acknowledged my presence in Les Diablerets, which had saddened me as she was widely regarded as an extremely clever scientist and was very popular with her colleagues. She was a pretty girl, not much older than me.

'Hello Amber,' she said. I was amazed and delighted she remembered me. 'I met you before,' she added, wrongly reading the reason behind my expression of surprise.

'Yes, in Switzerland,' I volunteered.

'That's right.' She smiled. 'I remember you.'

She walked up behind the reception desk and played with some files on the desk.

'Where are you working now?' I asked and pointed at the letters pinned to her chest. In Les Diablerets she had already been promoted from the Home Office to the Ministry of Defence and everyone said she was destined for high things. I assumed that this meant a lucrative job in private industry.

'Oh,' she fumbled with the badge awkwardly, 'uh.' She held her breath, darting her eyes back and forth between my face and her badge before mumbling modestly, 'CPNI.'

'What's that? I'm sorry,' I said, thinking that this was probably a famously cushy place to work and she'd be gutted I didn't recognise it, having often made this *faux pas* with solicitor acquaintances of mine. It probably had its own swimming pool. 'I'm sure I should have heard of it but I haven't.'

'It's uh.' She looked to the receptionist. 'Just trying to explain this.' She let out a laugh and turned back to face me. 'It's, uh, difficult to explain.'

'Oh right. Is it a new department?'

She cleared her throat and thrust some documents into a box. 'I don't want to give too much away. But yeah, security services. Thames House, basically.' She smiled meekly.

'Congratulations,' I said, not knowing where Thames House was but guessing it was some kind of special hi-tech police headquarters.

She cupped her hands under her waist, shrugged her shoulders and smiled. 'Thanks.'

An elderly man with sparkling eyes arrived to show me to my room. I wouldn't have guessed from his porter-like appearance and kind manners that he was Her Majesty's Prison Service Governor of Prison Dogs. Once inside the privacy of my room, I quickly freshened up and headed back out across the lawns separating the sleeping quarters from the castle in search of any stray drinkers. I'd missed dinner and pre-dinner drinks and didn't expect many people to be in the bar.

I followed the directions given to me by the receptionist. The long hallway was lined with truncheons arranged decoratively down the walls. Dating back from the nineteenth century, they varied in size and colour and I might have lingered over them longer had I not heard the sounds of celebration floating down the corridors. I turned left down the hallway and through the door to the bar. It was packed. There must have been at least a hundred men and a handful of women enthusiastically drinking. They spilled out of the bar area, into the adjoining games room and library. I drifted sheepishly from room to room, despairing at the lack of familiar faces, before returning to the bar. I'd quickly scanned the brochure given to me by the receptionist. Aeroflot, the Russians who had interbred dogs with jackals for the ultimate 'psychological weapon', would be present. International experts on the science of scent detection, whose work I'd read, would be here. And men from the olfaction department of the MoD would be making presentations. People stood drinking beer and whisky in small groups. I ordered a beer and took a seat alone. Sipping it nervously, I took a deep breath and turned to examine the people at the table next to me more closely.

It was Ben. I was so pleased to see him. Our eyes met and he failed to recognise me. I got up and went over to him and tapped him on the shoulder.

'Hello.'

He got to his feet and we kissed each other on both cheeks. I don't think we had done that before, but it seemed natural.

'How are you?' he asked, towering over me.

'Pleased to see you.' I smiled up at him.

'It's really nice to see you. Sorry I haven't replied to your emails. I've been so busy.'

'Sorry for asking you so many questions, and for banging on about odour conditioning, and everything else that I couldn't find in the library,' I said. 'I was just curious about sniffer plants the other day, and I didn't know who else to ask.'

'Well, we have been researching plants recently. There are some that naturally react to explosive material. Do you know Toby?' He pointed down towards the curly head of a man sitting hunched over the table. His profile was young and arguably handsome and he looked bored by the conversation of the darkly clad men opposite him.

'No,' I said and we both turned to face him again. He spotted us smiling towards him, and got up to join us.

'Amber, Toby. Toby, Amber,' Ben introduced us.

'You look familiar,' Toby said.

'Amber is researching dogs.' Ben said to Toby.

'I've been meaning to ask you,' Toby said, looking up at Ben. 'How many pesticide dogs do we have?'

'Pesticide sniffing dogs?' I said incredulously, suspecting a dog PR job. This seemed an odd time for Toby to have posed the question and I'd already decided that the media was being regularly fed fluffy applications of sniffer animals; if they weren't rescuing bodies in the wake of earthquakes, they were detecting cancer in loved ones' breasts.

'Oh yes,' Ben said to me, 'they're really good at detecting illegal pesticides,' before turning to Toby. 'We need more of them,' he said, crossing his arms and steadying himself, his face as ever a picture of earnestness.

'I didn't know illegal pesticides were a problem,' I said. My knowledge of agriculture is next to nil. 'Are there a lot of illegal pesticides?'

'Their sale has been effectively prohibited but some farmers stockpiled them and still use them in making our food, because they work; it's just that they poison humans.'

'Although I think we are still one of the biggest exporters of some of those pesticides aren't we?' said Toby.

'Yes.' Ben nodded gravely.

'Got to wonder about the ethics of that haven't you? Banning something in this country because of its toxicity but happily selling it to other countries?' Toby asked me.

'Yes,' I agreed. I liked the way he spoke and his green eyes were engaging.

Several passers-by had jostled past us by this point. The place was chock-a-block with the movers and shakers in Dog World. And the majority of them didn't look or sound like dog handlers. Toby suggested we sit at a nearby table.

'So how do you two know each other?' Toby asked as we closed in on the only free table in the room.

'Amber did a presentation for us at our conference,' Ben said, holding on to his knees as he sat on an armchair. I sat round the table next to him.

'And I told you about the smell conference,' Ben said looking over at me, 'and you came to that.'

'Yes. Thank you for telling me about that. It was really interesting.'

'Did I? I didn't tell you about this conference, did I?'

'No,' I said, 'you didn't.'

'The mind games at that other table were doing my head in,' Toby said, joining us at the table.

'I am new at this game,' Toby explained, 'and I know that there are some things we are not meant to reveal to security services from other countries but I can never remember what. '

'I am sorry to be naïve,' I said, immediately blowing any cover he might have assumed for me, 'but what sort of information couldn't you tell other security services about your research into canine olfaction?'

'Well, what we are using it for and what our best techniques are. Otherwise they'll nick 'em.'

'So,' I asked, charging my voice with national concern, 'what are we going to come up with to match the sniffer sharks the US plans on using to detect submarines?'

'Well, I had a good idea a few years ago but couldn't get the funding for it,' Ben answered.

'What is that?'

'Sabotage salmon.'

'Sounds good but I'm not sure that they'll be much of a match for the sharks,' I said.

'But they have the best olfactory system and it's the easiest to re-code. Basically salmon, when they are small and living in the river, encode the odours of their environment so that they will recognise them on their return from the sea to mate in the river. All we need to do is impregnate their water with the odour of explosives and then inject them with the right hormones when we want them on duty.'

'But the sharks will just eat them,' I said crossly.

'Well we can protect them with a fleet of dolphins. Toby's got his own idea though.'

'Yes. Electric eels. They are huge,' he said stretching out his arms to demonstrate, 'they grow to over eight feet.'

'Sounds promising,' I said.

'Yes, but how would you train them?' Ben laughed.

'Oh come on, that's easy. Stick them in a tank. They emit an electrical charge,' Toby puffed up his face and imitated the noise it made, 'when you prod them. All you have to do is emit the odours you want it to detect every time you prod it. Then stick a light bulb to it and send it off.'

'I think the light bulb will just blow!' Ben laughed.

'But even if it doesn't,' I said, 'you'd have to be very close to the eels to spot the signal.'

'Well, OK, large spotlights.' Toby flicked his head back, unruly curls of his tousled black hair immediately returning to his temples, where they hung in an unmilitary fashion. He looked like he'd be more at home in a jungle than a military compound.

'What about squid?' Ben laughed.

'Brilliant,' I said. 'Giant squid. They'd be perfect because of the mythology surrounding them.'

'Absolutely. The advertising campaign would have to feature Sea Monsters!'

'OK,' Toby interjected, 'I am seeing one submarine surrounded by a fleet of salmon with dolphin re-enforcements and further along another submarine with a load of sharks around it. Another followed by a load of squid or prawns or whatever.'

'It's chaos.' Ben laughed. 'But the crazy thing, Amber, is that this is what our meetings are actually like.'

'They start off very serious,' Toby added, 'but they end up with us coming out with all of these far-fetched ideas.'

'And sometimes someone somewhere funds them.'

'And as a result the world is a safer place,' Toby said in mock seriousness. 'Except for the sea, that is. Bloody nightmare out there.'

Toby left the table to buy us all more drinks, which we drank. And Ben bought us more, which we drank. The bar was heaving when my turn came to visit it and I had to wait among the throbbing throng of thirsty officialdom before I could get my order in.

A tall man in his early forties and wearing a dark navy suit said hello. His badge had the letters CPNI on it. His greying hair receded around his temples in a gentlemanly fashion. His eyes were softly cynical and his dimpled cheeks betrayed a wry amusement with life.

'I'm Alistair. I don't believe we've met. Though naturally I've heard of you.'

I was too tipsy to be anything other than flattered. He persuaded me to order a malt whisky, and once our drinks were served, helped me carry them to the table where he sat himself on the empty chair. He knew Ben and Toby.

'What do the letters stand for?' I asked him.

'Centre for the Protection of National Infrastructure.'

'Thames House?'

He smiled. 'We've moved actually.' He stretched his legs out. 'To a secret location.'

The night was young and the whisky in the subsidised police college bar was cheap and plentiful. It was 3 a.m. by the time I got to my room.

Elated, I phoned Cass in the States.

'Marjorie. You're drunk.'

'I have been drinking.'

'I thought you were on a Dogwatch mission?'

'I am!' I said triumphantly. 'I have been drinking with top dog men. MoD and security services zoologists.'

'They have zoologists working for the security services? Oh Jeez.'

'*Handsome* zoologists,' I hiccupped. 'One of them even took me to sit outside on a bench under the stars. I think defence must be all about zoology now. The chief scientific advisor to the Ministry of Defence, probably the most influential scientist in the country, is a zoologist.'

'Well, I'm glad you're back with dog people. That's where you seem happiest, Marjorie.'

'I've just spent two hours under the stars drinking wine on a bench in a landscaped garden with a rather good looking man from the security services, actually,' I slurred boastfully. Alistair and I had polished off the remains of the discarded bottles we'd found littering the empty bar.

'What's he doing at the dog conference?'

'Apparently,' I hiccupped again, 'this event is organised by the security services.'

'Honey, don't you think it's a *bit* weird that handsome security service guys are romancing you under the stars and getting you *drunk*?'

'No. Not at all. They're very nice. They are not *all* from the security services. They just organise it and so there's quite a few of them here. My favourite people are the military scientists. And I am romancing *them* and getting *them* drunk.'

'Oh, come on honey. How big are these guys?'

'Really quite big.' I laughed, their broad shoulders, firm chests and muscular arms, tightly buttoned in crisp white shirts, coming back into view.

'And you think you can drink them under the table?'

She had a point.

'And what are all these people from the security services doing at a friggin' dog conference?' She sighed. 'Did you *tell* them anything?' she asked.

'What do you mean?'

'Well you didn't tell them that you are thinking of writing about what they're doing or anything dumb like that?'

'Yes, I did actually.'

'Are you . . .' she said in a high-pitched tone before pausing and emphasising the word '*kidding*?'

'No. They seemed very pleased. They said they would help. I told them I was looking for animal stories.'

'And did they give you any?'

'One of them,' it had been Toby, 'told me a joke about a bear.'

'A joke about a bear?'

'Yeah, this bear comes out of the forest, it's starving, hasn't eaten for months, it's been in hibernation.'

'Uh-huh.'

'And it goes into a lodge and there is a can of food on the table.'

'Yeah.'

'And the label reads "spam".'

'Like the stuff in the Monty Python sketch?'

'Exactly.'

'And?'

'And the bear doesn't eat it. It is so clever that it could *smell* that the tin contained spam.'

'That's it? That's the friggin' joke? This is how you've got them to spill the beans? To tell you a bad joke about a bear?'

Cass didn't get it. She couldn't see that this had been Toby's way of telling me that the whole odour detection thing was a ruse.

'Any other *animal* stories?' she asked.

'Yeah. I was speaking to a guy from the US military who told me about these dolphins the navy have that can handcuff people and attach them to a line so that the navy can trawl them up.'

'And you'd told this guy you were writing a book too?'

'Well, word got around. He knew.'

'And do you think he might have been having you on? Or feeding you horse-shit?'

'No.'

'You think the US navy has trained dolphins to arrest people?'

'Yes. Why not? I've read about their sentry dolphins – trained to recognise Navy Frogmen – mistakenly attacking and killing them. They have an undersea warfare department.'

'OK, but arresting people? How, honey? It's not physically viable.'

'I asked him about that. He explained. He,' I hiccupped again, 'got me by the arm and explained about this bar they use.'

'Hmnn.'

'There was one thing I regret telling them,' I said, feeling more sober now.

'Really?' she drolled sarcastically. 'Worse than blowing your cover? What?'

'I told them I was scared of spiders.'

'Honey, a lot of people are scared of spiders.'

'But they told me about these really big ones in Iraq. They are huge and they are scared of sunlight and so they hunt for shade and out in the desert, *you are the shade*. And so the spiders come up to your feet and when you run they run after you and they are *huge*. They are scarier than insurgents.'

'Insurgents? Honey, you've lost all perspective. The MoD is not going to chase after you with spiders. There is no room 101. You fucked up in telling them about the book. Now go to bed, lil' chicken and sober up.'

'Cass, are you by a computer?'

'Yes.'

'Can you look up CPNI for me?'

'Sure,' she typed away. 'It's part of MI5, Marjorie.'

I said goodnight, promised not to drink so much tomorrow and climbed into bed. My head was swimming with tail-ends of the night's conversation. Despite being 'inundated' with gun dogs 'in the aftermath of the hunting ban', the British police force was now running low in its supply of them. The US was buying up all the dogs in Europe, 'they're paying up to £10,000 for a German shepherd. And the breeders know they can get that kind of money now.' I thought about Cobra, the civil contingencies committee, and what an interesting name that was for our country's emergency response team. The security services seemed to have a fetish for animals. Alistair had told me his departmental head had threatened to resign if any more animals were recommended for security duties. 'I can tell you that,' he'd said, 'because he's about to resign anyway.' He'd also told me about Viet Cong guerrillas washing in American soap to avoid detection by their dogs during the Vietnam War. Toby had told me how his zoology studies had mainly consisted of watching David Attenborough's documentaries. I'd seen the pleasure David Attenborough derived from illustrating how much cleverer nature was than modern technology, regularly comparing the aeronautical capabilities of the helicopter to that of various flying insects. And I expect it riled the military. In the US whole new disciplines have been created in what could be interpreted as an attempt by its military to earn their stripes in Attenborough's eyes. 'Biosystems Research' looks at how organisms can be used to support military operations. 'Biohybrid Studies' devises technologies to exploit natural abilities and instincts – like attaching cameras to flying insects. 'Biomimetic Programmes' mimic nature by creating machines such as the robotic sniffer lobster. I'd read enough quotes from military representatives about the inspiration they took from biology to know what the future now held in store for us. A giant octopus floated into my mind; an obvious candidate for marine intelligence. They can escape from locked tanks, break into other tanks and kidnap lobsters; renowned masters of disguise.

CHAPTER TWENTY NINE

The Leash Men

They set a slamhound on Turner's trail in New Dehli, slotted it to his pheromones and the colour of his hair. It caught up with him on a street called Chandni Chauk and came scrambling for his rented BMW through a forest of bare brown legs and pedicab tyres. Its core was a kilogramme of recrystallized hexogene and flaked TNT.

William Gibson, *Count Zero*

The morning's talk on the detection of corpses did little to assuage my hangover. Forensic scientists bemoaned chemists' failure to synthesise an accurate representation of the scent of human death. The pseudo scents that were available on the market were useless for training cadaver dogs on. Pseudo scents, I had already discovered, were big business. It was possible to purchase everything from the scent of fear to the scent of death. And cadaver stench systems were already being promoted as less-than-lethal weapons at security exhibitions. The scent of death was not sufficiently refined for the dog detectors, however. Consisting of cadaverine and putrescine, they emitted the smell of carrion accurately enough, but not that particular aroma that was unique to the odour of decomposing human flesh.

During the coffee-break I got talking to a young woman dog trainer, who worked for a private dog detective agency and specialised in the detection of dead bodies.

'Do you use pseudo scents as training aids?' I asked.

'No, 'cos like he said, they're useless. The dogs would just pick up on any dead stuff.'

'What do you use?'

'Pig skin. Closest smell we've got to human skin.'

'Oh right.'

'It's a bit grim actually. I have to bury the pigs, 'cos we need them at different stages of decomposition for training the dogs on.'

'You have to bury them yourself?' I asked, amazed by the thought of this pretty young woman digging graves to bury pigs in.

'Yeah, unfortunately. I don't have to kill them though, which I am grateful for.'

'Must be rough,' I sympathised. She was about my age and seemed very amiable.

'Yeah and I mean pig skin's not the best but we are not allowed access to human remains in this country. I'm always asking my dentist for decaying teeth – I don't think he thinks it's what a woman my age should be interested in though!' She laughed. 'But we don't have a body farm in the UK,' she shrugged.

'A body farm? What do you mean?'

'Oh haven't you heard of it? It's quite famous. It's been on TV.'

'No.'

'It's this place in the States. People donate their bodies to it instead of to medical science. It's for forensic research. It's amazing – they've got bodies hanging out of car windows, trapped in car boots, the lot.'

A senior police dog handler I'd seen speaking to Geraldine last night came over. I'd been concerned when I'd spotted Geraldine draped over the arm rest of a sofa at around midnight. I'd gone up and offered to buy her a drink and then got waylaid by a man from the security services. I'd feared Geraldine was out of her depth in the company we were keeping at this conference. She'd no doubt counted on chaperoning me but as she sat drinking whisky with dog handlers from Northern Ireland, I was laughing and drinking with a whole team of MoD zoologists, none of whom, to my surprise, had heard of Geraldine. But in fact, she'd told me at breakfast, she'd had a whale of a time, catching up with old friends and dog handlers. I'd not met any handlers apart from the couple with the attack dog on my way in. Everyone I'd spoken with had been working for the MoD, MI5 or the Home Office Scientific Development Branch.

'We've been talking about whether we could have one in the UK,' the senior police dog handler said. 'Whether there'd be public support for it.'

'There might be,' I said. 'If you made televised requests for donations after *CSI*.'

'Yes. Even just people having leg amputations. They just throw them in the bin. It's such a waste.'

'Yes,' I nodded, wondering how common amputations were. 'What do you train CSI blood dogs on at the moment then, pseudo scents?'

'Excuse me,' the woman from the private dog detection agency said. 'I'm just going to get some cake.'

'Oh no,' he said. 'As you've heard, they don't work. We,' his face heaved itself up for a smug smile but dropped again suddenly. 'Don't get me wrong, we don't go round illegally pinching body parts.' I remembered something a solicitor had told me, about a dog handler bragging that he could get hold of blood and dead body parts. 'We train them on blood as well.'

'Where do you get that from?'

'Well, our employees could give it to us. I don't think there'd be anything wrong with that.'

A strange expression must have come over my face because he clearly felt the need to elucidate this option despite the fact that this didn't appear to be one they presently used. 'There aren't any regulations against that. So long as they've consented. I'm sure they'd be happy to. I'd give my blood to train my dog on.'

I didn't know what to say. This conversation had gotten bizarre.

'And it might sell,' he burbled on. 'The public might be up for donating their bodies to forensic science.'

'Yes. Yes. They might,' I said, nodding my head vigorously.

'So,' he said. 'Geraldine tells me that you used to be a lawyer.'

'Yes, that's right.'

'I know there are lawyers out there, arguing that a sniff is a search, but it's not. The dog is picking up on molecules in the surrounding air. But I'm sure you're aware,' he smirked, not bothering to elaborate further on this fine distinction, 'of the technical argument.'

'Yes,' I said, 'I've read the Association of Chief Police Officers' guidance on their use. It seems to me that some police practices might be in breach of it.'

He raised his eyebrows expectantly.

'Police aren't meant to force people into being sniffed are they? I think the guidance uses the word "funnelling". People aren't meant to be "funnelled" past them.' He nodded in agreement. 'But,' I continued, 'it seems to me that's precisely what the police are doing when they place one at the top of an escalator.'

'Well,' he simpered. 'We don't have any power to tell people to walk past it, but we can take advantage of the natural environment.' I thought this was a strange term for escalators.

'And we don't just take our drug dogs out randomly. These are intelligence-led operations.'

'Really? I thought they were just taken to train stations to sniff all passers-by?'

'Yes, but we have intelligence that huge numbers of people who take drugs use the transport system. And to be honest, we don't often get complaints.'

'You've had lots of complaints of dog bites though haven't you?'

'Yes, but then we are talking about a different kind of dog.' He was right.

I told him about the 73 per cent failure rate they'd found in Australia and asked him what he thought.

'Well,' he said. 'That's the thing. You can't really measure it in terms of the number of positive finds. Let's say, for example, that you had close associations with a well-known drug dealer.'

'OK,' I nodded.

'We would know that from our intelligence sources. And so we might approach you with a dog. Just because a dog gives us an indication doesn't mean,' he said looking down at my crotch, 'that we could look in your knickers. We can only conduct a cursory examination as a result of an indication. So the dog could well have been right, but there's nothing we can do about it.'

'Sure,' I said, ignoring his gratuitous reference to my underwear. It was a habit many security men seemed to have. 'So lots of the time the dog might be right, but you're unable to prove it.'

'That's the first point,' he said. 'Then there's the fact that admissions flow out of people indicated like a gush of air – they're so relieved not to have anything on them. We keep intelligence reports on incidents, when we have the time. Sometimes you can't obviously – just because of the sheer number of indications.

'How many indications did you get at that festival last week?' he called over to a passing colleague who was on his way into the next talk.

'Seven hundred and thirty-five.'

'Gosh,' I said.

'I mean, when we're dealing with that number, we can't be writing a report on each one. But we make them when we can. So that when we get criticised – and that time may well come . . .'

'Like it has,' I interjected, 'in America. There have been hundreds of challenges to the legality of sniffer dogs there. Many of them successful.'

'When they ask,' he continued and I wondered who 'they' would be, 'we'll be able show them with these records that it's not just as simple as drug finds and no drug finds. A lot of these people take drugs, or go to places where drugs are consumed. If they've told us that they were drinking in a certain establishment, and used that fact to explain the smell of drugs on them, we can visit those premises and find out what's going on there.'

I thought about informing him of the Italian study a forensic toxicologist had told me about, in which random samples of the air above buildings had been found to contain traces of cocaine, but I didn't want to antagonise him.

'Yes,' I said. 'The dog is a very useful intelligence tool isn't it?'

'Ion-track scanners are good too. Some of my colleagues have been asking people to volunteer for drug swabs in pubs – and it's amazing – they all do it – including one guy who actually had drugs on him. It's like "Who's a mug day"!'

I laughed encouragingly.

'But the dogs are unbelievable. My dog can spot someone with drugs on them from fifteen feet away. I dunno how he does it. I don't think we have any idea about the extent of their capabilities. People are saying now that they might have infra-red senses.'

'Are they? Where do they get that from?'

'Research on wolves. And people talk – don't they – about auras? People having auras. Maybe there's something in that? I mean just 'cos *we* can't see it – doesn't mean it isn't there. Of course until dogs learn to talk – we'll never know.'

I was beginning to see why dog handlers would enjoy this conference, with all its scientific talks to inform the boys in blue back home about. Dog men are in a class of their own among the police force. A law unto themselves.

'So,' he asked, 'have you seen many of our dogs in action?'

'Yes,' I said, 'but I'm not planning on writing up all of my research,' I tried to reassure him.

'No,' he nodded, 'you wouldn't want to. If you were to say anything critical, I'd get a crew of dog handlers to stand outside your house and howl.'

'That sounds nice,' I said assuming he was pulling my leg.

'It won't after three weeks,' he warned as I followed him into the auditorium.

The next presentation seemed to echo the handler's sentiments about the mug's response to police surveillance. A high-ranking police officer gave a run through of operations conducted in the 'leafy suburb' in which he operated. His force had been using dogs and ion-scanners round nightclubs to great effect.

'These days people go straight to the machine with their hands up. Some even recognise the dog from school demonstrations we started doing a few years ago. The message is obviously getting through as to what they can expect from the night-time economy.'

He laughed at the exaggerated portrayal of drug dogs' abilities in street addicts' minds. 'They think dogs can detect them from 300 metres away!' I wondered what life must be like for the poverty stricken heroin junkies in this man's leafy suburb and whether they'd have heard of the moths and the bees. Words from Genesis flew through my head: ' . . . the fear of you and the dread of you all shall be upon every beast of the earth and upon every fowl of the air.' There was, the officer explained, 'a lot of paranoia in the addict community'.

Delighted as he was with the impact of the dog on the drug-using community, it wasn't necessarily his favourite tactic. It seemed that his toy of choice was the ion-track machine. His force had successfully closed licensed premises on the basis of its readings. It had refuted claims for damage to houses they'd busted into by showing that although no drugs had been found there, their machines registered the presence of drug traces, proving that the police had been correct in their suspicions. The ion-track machine, he said, might be expensive but it had saved the police force a large amount of money in civil law suits, where, he reminded the audience, the burden of proof was the 'balance of probabilities' instead of the less easily obtainable criminal standard of 'beyond reasonable doubt'.

He felt his force had done rather well at fostering community relations among law-abiding households. They'd been conducting 'house-to-house' enquiries and found that citizens were becoming more forthcoming in the information they provide to the police. He thought this was as a result of the normalisation of these informal chats, 'they no longer feel like a grass'.

I had a word with the speaker on our way for a group photograph later on in the afternoon. He told me his force had a criminal case coming up, involving a 'Mr Big' and they were going to try to rely on ion-track readings in it. The Crown Prosecution Service was up for giving it a go. He became unnerved when I questioned him further, wanting to know my credentials. I told him I didn't have any.

By the end of the day's talks, I was in need of a stiff drink. During one of the presentations, an electronic nose designer and salesman, who explained how his e-nose could sniff milk and 'tell you which cow it's come from', had mentioned the need to address ethical considerations when developing this technology and I'd told him at lunch that I'd like to discuss what he meant. I hoped to find him in the bar and kept an eye out for him as I mingled among the peculiar mix of low-ranking dog handlers, senior dog policy advisors, scientists and security services from around the globe.

'Not just searches that we do,' a Welsh dog handler told me in the bar's back garden where I'd overheard him grumbling about 'the people who talk in isms'. 'Sniffer yeast?' I'd heard him say. He was commenting to his colleagues on a talk we'd heard that morning about 'biological alternatives' to dogs including 'artificial immune approaches for anomaly detection' that could 'mimic the immune system's response to foreign substances'. Biological cells, the scientist had explained to us, are 'self-contained sensors' and yeast cells are a particularly robust and 'tough beast' for the job of target detection. 'There are no psychological issues like those we've heard of with the dog,' he'd explained, provoking an irritable murmur of dissent among some of the audience. 'The cells are "readily manipulated",' he said before concluding with the prospect of speeding up millions of years of evolution in 'just a few weeks'.

'Sniffer yeast?' asked the dog handler. 'How are we going to use that? Not very practical is it? Wandering round with bloody pots of the stuff, and a hive in the other hand.' I could picture his ungainly body struggling to balance the two in his arms as he patrolled the land, a disheartened Winnie the Pooh in an unfriendly forest of unpredictable enemies.

I'd introduced myself to this nonchalant Welshman with a story from an article I'd read about a police officer being convicted in Nigeria of trying to pass his sheep off as a sniffer dog, before hastily reassuring him that I myself was Welsh and no personal insult had been intended by this account.

A working-class copper from the Welsh valleys, he'd applied for a job as a dog handler several years ago. Unlike a number of handlers I'd met, I didn't get the impression that he liked dogs. 'No brain at all. Thick as a goldfish,' he'd said when I asked him what he thought of the morning talk on canine cognition. The speaker had rebuked the police force for 'under-using the dog's abilities' and explained how olfactory lessons were probably the first lessons in behaviour learned by any brain. Dogs, she'd said, to the consternation of most members of the audience, are the most similar to the human in cognitive terms, 'even closer than chimpanzees, because we co-evolved'. In sharing, our human society, she'd said, the 'gene expression in the dog brain' had adapted accordingly and shed its wolf's clothing. She said that a dog brain achieves 'categorical thinking' or 'fast mapping', which, she said, should be a cause of concern to handlers because this is 'the first stage in language acquisition'.

'Bonkers,' said the Welshman. 'Doesn't understand a bloody word I say.'

Adee Schoon, probably the world expert on the science of scent detection, had told me the challenge she faced persuading dog handlers of the importance in eliminating cues a dog could pick up on, like different coloured labels on the jar containing the suspect's scent in a scent line-up. Despite emphasising the need to ensure that the handler in a scent line-up is unaware of which sample belongs to the suspect (because of the danger of communicating their suspicions to the extra-sensory perception of the dogs), handlers in Schoon's native Netherlands were facing charges for failing to adhere to this police protocol.

The Welshman told me about a training exercise he'd undergone with a previous dog 'a few years back'. The Welshman's trainer had obtained permission to use a local school as a training ground for the day, 'one of those places for rich Catholic girls, on a working farm,' he said. ''Orrible place it was; pigs, cows, the lot.' He rested his leg on one of the wide stone steps descending from the police library room, furnished with salon-style coffee tables and chairs and filled to the brim with drinkers spilling out from the adjoining bar room.

'Not the best place to take a dog,' he continued. 'Especially one like the psychotic bastard I had at the time.'

'No,' I agreed, taking a sip of beer from my glass.

'Anyhow, instructor tells me to introduce my dog to the cows.' He sighed resignedly. 'I said "You must be kidding." "Go on," he says. "Otherwise they'll be distracted by them." "All right," I say, so I took my dog up to one of the cows. "'Ello Cow," I says to the dog. Well, it only clamps its jaws on the cow's nose, doesn't it? "Moo," it's going.'

He was telling me the story to amuse me, and I laughed, clamping a hand over my mouth in horror.

'Well,' he said, looking down at the pebbles on the ground. 'It breaks my heart to say this, but I 'ad to beat it just to get it off the cow. Instructor said, "Right, let's go."

We were like a bunch of naughty schoolboys. We got into the van and drove off, leaving the cow there with blood spurting out of its nostrils.'

I began to feel sorry for him as he outlined the hardships of his job. 'Other sectors of the police force see us as a bit daft 'cos we work with animals,' he'd told me. 'It's not just searches we do,' he said. 'A lot of it is wandering up and down in yellow vests with dogs. Trying to deter terrorists,' he scoffed at the futility of his working days. 'Mostly it's people coming up to us and petting the dog.' He took a sip of his lager. 'Adds something different to commuters' lives, I s'pose. They get to see the police in a new light.'

A couple of hours later, still unable to find the e-nose man, I resumed my conversation with Alistair, about the growth of security in the UK.

'We take terrorist incidents too seriously,' was how he'd started our conversation. 'Make ourselves look more vulnerable than we are. The number of people killed in the July bombings compared to the number of people killed in car accidents is minimal. I sometimes wonder what would happen if we just shrugged our shoulders when terrorist incidents happened.'

'Have you suggested that the security services experiment with chilling out about them?'

He laughed.

'Or are you suggesting we worry as much about car accidents and protect ourselves from them in a similar fashion? Your point could go either way.'

He laughed again and went to fetch us more drinks.

Several beers and whiskies later, we were sitting outside in the stairwell, where I tempted him into smoking a cigarette. After discussing the fallibilities of dog searches, and how I'd feel if the dog were infallible and zero tolerance became a realistic option (we'd need a full scale revision of all offences to ensure they merited outright prohibition) he reminded me that the dog's principal purpose was as a deterrent.

'I know,' I said. 'But how far are you comfortable in taking deterrence? I think the interesting question raised by technologies of control is that it forces us to question these issues. The deterrent value of a police officer on the street was incidental and taken for granted as a good thing. New technologies will permit you to push deterrence to its extreme.'

'Like the mosquito repellent?' he suggested.

'Yes,' I said. 'That's a good example.' The Mosquito™ is a device that has been installed by a number of private business premises. It emits a high-pitched buzzing sound over a range of fifteen to twenty metres that only teenagers can hear and is designed to deter them from congregating in a particular area, a pastime viewed as anti-social in Mustapha Mond's *Brave New World*.

'You know that the technology behind that was first experimented with on animals?' he said. This didn't come as much of surprise. I thought about the references to cattle-prods at the less-than-lethal weapons symposium, and the growing use of taser guns.

I asked him his opinion on the encroachment of Big Brother.

'I don't have an Oyster card,' he said. 'That's all an Oyster card is – a radio frequency identification chip. I don't want my every movement to be tracked.'

I nodded drunkenly.

'You know it is much, much cheaper to get an Oyster card? I mean you are essentially being blackmailed into getting one?' he said.

'I know. Though you don't have to register the card to your name and address.'

'That's right.' He nodded.

'But most people do so for the convenience of topping it up at home.'

'That's right.' He laughed.

'Thing is,' he said. 'People bang on a lot about Big Brother knowing everything. Big Brother isn't actually that clever.'

'I'm learning that. But I don't find that very reassuring 'cos he might act on his information even though it's unreliable.'

'That's true.'

We talked at length about the number of cameras and listening devices being sewn into our infrastructure.

'The thing is Amber, we are not individuals. None of us is.'

'What do you mean?' I asked, surprised by this sentiment. He'd given me the impression that he didn't like the secure world into which we were headed.

'We can't work without paying tax. We can't travel without a passport. We can't opt out of the state.'

'I could run away to the countryside and grow my own vegetables,' I said defiantly, thinking about Werner and the flowers that hung from his bed in the garden. 'I know when it is time for me to move inside, because they droop,' he'd told me, outlining his seasonal sleeping habits while we munched on the flowers in question from our salad bowls.

'European satellites would monitor you to see how many potatoes you grow!' He laughed.

'Don't be silly.'

'Yes, that is what they do.'

'Well I could nick other people's potatoes. I could become a homeless drunk.'

'Yes but they'd find out.'

'Fine. But why does that stop me from being an individual?'

'Because you don't have a choice about how you live your life. Because we are all part of a larger organism.'

'Like a hive or an ant colony?'

'Sure.'

'But I am still an individual. I can still rebel. It's a relative scale.'

'Yes.'

'So at what point are you saying that we cease to be individuals and become agents of the state?'

'I don't know. It's an interesting question. The trick is,' he said, as if about to suggest a means of getting off the barbed wire path we were travelling down, 'to learn from the Romans.'

'What do you mean?' I asked, my addled mind conjuring up images of lions' dens.

'Keep people happy with cheap entertainment.'

'They do that in Spain with cheap fags and booze. What's our method? Shopping centres?'

He just smiled and nodded. We were both flagging.

Getting back to his obsession with Rome, he soldiered on into the late hours, trying to articulate the differences between safety and security, how the two weren't the same thing and how his work was a constant struggle to find the right balance between the two.

'Well,' I said, 'it's been great talking to you but I think it's time I went to bed.'

'Yes,' he said looking up at the brightening sky. 'I should really go back to my room now and write up my meeting with you.'

'Really?' I laughed.

'Just joking. It's interesting how far paranoia can go.'

'Oh,' I said as we traipsed back through the castle. 'Do you know how I could contact Eliza Manningham-Buller?'

He looked at me in surprise.

'I just want to interview her about bees.'

'Bees?' he asked, leading me out the other side of the castle and on to the paths to our respective bedrooms.

'Yes, I understand she keeps bees.'

'And sheep,' he said, suggesting that they were on intimate terms.

'Sheep?'

'Yes, our co-director is going to do the same thing when he resigns. You can keep both with a relatively small amount of land.'

They sounded like characters out of Philip K. Dick's *Do Androids Dream of Electric Sheep?*

Outside my room, I thanked him for walking me there and bade him goodnight.

The next morning, as I waited for my taxi in the car park, the electronic nose man spotted me from within the castle's walls and joined me outside. His eyes hovered over my shoulder as we spoke. A couple of dark-suited men, of Middle Eastern appearance, walked past us into the site of the conference talks.

'No one knows who those men are or how much we should be revealing about our research. This is all highly sensitive information.' I'd asked him yesterday if I could have a word with him about the ethical issues raised by his research. He was the first person in the field of odour detection to have mentioned them of his own accord in my presence. He had done so in reference to his discovery that he could detect at what stage in her menstrual cycle a woman was at, just by sniffing at her from the other side of the room with his electronic nose. It was called stand-off detection and in its performance of this particular function, he described it as the 'Vatican's nose'. I wanted to find out what, if any, other ethical qualms he had about his work.

He'd said we'd catch up later and warned me that his ethics were Christian in nature. I was flummoxed as to what a Christian take on olfactory surveillance could be and was pleased he'd come to speak to me before I left.

'So many ethical issues. I mean we are talking about decoding your genetic make-up, your immune system. Pretty useful information to have on your enemy. We're not there yet. But we will be soon. You know, in the airport they get you to walk through machines that blow particles off you for analysis. You sign some sort of disclaimer. What do we know about what they are doing with it? It raises data protection issues and all sorts.'

'Well yes. Surreptitious stand-off detection of diseases would raise a number of ethical issues.'

'Or *vulnerability to* disease.'

'The woman speaker on canine cognition yesterday suggested dogs could be used as "bio-markers" to detect schizophrenics, the implication being that they could be kept out of safe neighbourhoods.'

'AIDS, TB, you name it. I am working on STD detection at the moment. Stand-off detection of them.'

'You can detect sexually transmitted diseases?'

'And different races. Ethnic background.'

'Are you sure?'

'Absolutely. We were approached a few years ago by the South African police – wanting to know the odour signature of blacks.'

'So, what did you say?'

'Well, there was a different managing director then. He was an ethical man. He refused to assist them. But the War on Terror railroads any attempt to impose ethical standards. The rule of law has gone.'

'So, if you feel this way about odour research, that it is unethical, why are you in it?'

'Because I believe that our technology is the most

sophisticated odour detection device out there – and I want to make sure the UK has it rather than the US. The amount of funding going into e-nose technology in the US is astronomical. The companies go bust after a few years because the technology isn't ready for application yet, but that suits their purposes. The research is coming along. We'll be there in five years – we'll be able to track your ethnic origin and identity. And I don't like where all this is headed, but it won't stop. I'm less naïve than when I was younger. I've visited war-ridden countries. The world is an unpleasant place.'

'But the MoD told me that e-nose technology is a long way off and that they prefer the biological model?'

'Have you spoken to the security services about this at all?'

'No.'

'Well I have. I've been to their building. I didn't know that's where I was until they told me.'

'What is it that disturbs you most about military olfaction studies? What is your vision of the future?'

'Plagues of locusts,' he said cryptically.

'What do you mean?'

'Genetic weapons. Insects trained on target odours, sent out to attack.'

Genetically programmed plagues of locusts. It made sense. That was what bio-ethicists were getting at in their papers on biological and chemical warfare. I could sense it all. Butterflies netting airborne confessions for laboratory analyses. Cultivated slime mould conditioning our neural networks with volatile messages. The smell of titan arum, the smelliest plant known to man, pumping out volatiles of decomposing flesh and activists fainting at its feet. Bombardier beetles blasting protestors with chemical sprays until their skin is scorched with blisters. Sharks gobbling foreign-scented swimmers. Box jellyfish jet-propelling themselves after underwater Greenpeace divers quietly monitored by a colossal squid eye. Smart sand dusting our feet as we walk along the beach. Children will be taught to respect Mother Nature.

'What about odour influencing?' I asked him.

'Yes, you can affect people's behaviour with odours, especially if you combine them with sound.' The future was becoming clearer. A biochemist seeped into view, silently weaving neural handcuffs from the cobwebs of our minds. The bugs, her familiars, equipped with novel odours to give and find, arrest us with the memories which her clever scents unwind.

'How could you use odour to influence people?' I asked.

'Have you experienced a wave of fear go through a crowd? I mean, I believe in demonic entities. I think the Devil is on the rampage at present and only angels can distract him from his path. I am a religious man. But I wouldn't be surprised to find that odour was the means by which the sensation of fear is spread among a crowd. That's how it could be used.'

This conference hadn't been about dogs. The place was crawling with secret services and everyone was gossiping about the Arabic contingent who'd arrived yesterday morning, none of whom were wearing badges. The Russians weren't there to show off their canine prowess. They were there to find out information. Scraps of the day chased each other rabidly round my mind as I tried to piece them into a meaningful context.

'Vhat is the smallest amount of explosive you can detect?' the long-blonde-haired Russian had asked an MoD speaker. 'I can't say,' he'd replied. This conference was about espionage. Sure they had sprinkled the audience with uniformed police dog handlers who didn't have a clue about the science they were hearing. And they'd invited the woman from Cancer Detection Dogs to look worthy. But it was all a front. I glanced down at the crystal glass laser-etched cocker spaniel key fob that I'd found in my conference goody-bag. The dog really was the public face of the war on terror.

Squirrel Spy Ring? Thats Nuts!

Sky news website
12 July 2007

Police in Iran are reported to have taken 14 squirrels into custody – because they are suspected of spying.

The rodents were found near the Iranian border allegedly equipped with eavesdropping devices.

The reports have come from the official Islamic Republic News Agency (IRNA).

When asked about the confiscation of the spy squirrels, the national police chief said: 'I have heard about it, but I do not have precise information.'

The IRNA said that the squirrels were kitted out by foreign intelligence services – but they were captured two weeks ago by police officers.

A Foreign Office source told Sky News: 'The story is nuts.'

But if true, this would not be the first time animals have been used to spy.

During World War II the Allied Forces used pigeons to fly vital intelligence out of occupied France.

More recently, US marines stationed in Kuwait have used chickens as a low-tech chemical detection system.

CHAPTER THIRTY

Headcase

> ... *The men in silence, far behind,*
> *Conscious of game, the net unbind ...*
>   Jon Gay, *Plain Words to the Spaniel*

> *The little dogs and all,*
> *Trey, Blanch, and Sweet-heart, see, they bark at me.*
>   William Shakespeare, *King Lear*

Shami Chakrabarti, director of Liberty, recounts how the chief scientific advisor to the government summed up the erosion of privacy in the UK to her: 'If you put a frog into a pan of warm water, it will try to escape. But if you heat the water very slowly, the frog notices nothing and allows itself to be boiled to death.'

Life in the UK did seem to be changing. Transparent luggage was fashionable as were web confessions (or open-source intelligence as CID officers call them), and the clipping of large electronic tracking devices to one's ear lobes. Phone calls were no longer being recorded just for training purposes, but for security as well, and no one seemed to know what that meant. Interested third parties were remotely activating in-built microphones of personal technologies. Terahertz 'See through your clothes' scanners, now informally referred to as 'T-rays', were being operated round London from the back of converted vans without any legal basis and CCTV operators, when they weren't barking orders at passers-by, were taking teenagers to court for flashing their breasts at cameras in response to being followed down Brighton beach.

Someone had erected signs in tube stations informing us the police would be conducting random searches and that our cooperation would be appreciated. Small cars drove round with cameras on the tops of long posts. Microphones had been attached to CCTV, analysing the passing moods of voices below. Clubbers were giving their fingerprints and dates of birth in return for organised and closely monitored leisure. Practising lawyers debated digital evidence and the dangers of disguised doctoring. The *Independent* explained that 'new technology and "invisible" techniques' were being used to gather information on UK citizens. There wasn't much talk about what these invisible techniques might be. 'The level of surveillance will grow even further in the next ten years,' continued the *Independent*, 'which could result in a growing number of people being discriminated against and excluded from society.' Our medical records were undergoing computerization and proposals were being drawn up for making them available to international law enforcement agencies.

The security threat was getting bigger all the time. According to Sir Richard Dearlove, the former head of MI6, al-Qaeda, and its terrorists' cells, was 'showing an extraordinary ability to mutate'. And so were terrorism powers, numerous environmental protesters getting caught up in their ever-widening nets.

The police explained they were unable to foresee attempted bombings because there were too many suspects to investigate. They needed more resources. Surveillance helicopters and spy planes had been purchased and piloted.

With government accusing the police of police state tactics and the police accusing the government of Big Brother antics I didn't need the Information Commissioner to tell me what was going on, but I listened. We were 'waking up to a surveillance society all around us'. We were surrounded, the media was telling us. We could all see the cameras, and everyone was talking CCTV.

I asked a psychiatrist acquaintance of mine whether the widely noted prevalence of surveillance technology hampered his ability to diagnose paranoid schizophrenics whose predominant preoccupation was fear of being watched by the security services. Surveillance, he told me, was now a way of life, and if a person found that too difficult to deal with on a day-to-day basis, that might indicate that there was a mental problem.

I felt I knew how secure counter-terrorism hawks wanted our world to be and how tightly secure they planned to fasten us within it. I listened intently to bio-ethicists warning against the 'imminent militarisation of biology.' I watched legal academics discussing the usurpation of the rule of law by the rule of technology in an environment of ambient intelligence capable of overriding human volition. They feared for the future of moral autonomy in an automated world where mobile phones transform carriers into reluctant roving microphones at the behest of the intelligence services; nano-technology reaches into the past to record yesterday's conversations; technologies of 'human enhancement' weaponise our soldiers; and increasingly complex algorithms steer civilian behaviour. Artists dream up smart clothing for coating us in chemically enabled bio-sensors to dose us in accordance with desired behavioural outcomes. Fashion, it is said, 'makes technology acceptable' and clubbers in Spain are lining up to get micro-chipped for access to VIP areas. Signs have been erected on transport routes, telling us to 'Trust Your Senses' while less-than-lethal weapons manufacturers ready themselves for the deception of our perception.

I didn't need Tom to tell me the Surveillance Society had arrived. But I listened, over dinner, at his house.

'It's outrageous,' he said, strolling out of the kitchen in baggy grey tracksuit trousers and laying a pizza down on the table. 'I'm on the verge of taking up arms.'

'I've been thinking of stocking up on insect repellent and nose clips myself,' I said, omitting any reference to woollen underwear. I was learning to control what I said to him as I struggled to avoid eye contact with the man in the bird-beaked mask. The masked man was patrolling my mind, stalking its darker recesses whilst promising me protection from everything he could find.

Select Bibliography

Smell

Ache, B. W. and Young, J. M., 'Olfaction: Diverse Species, Conserved Principles', *Neuron*, 48 (Nov. 2005), 417–430

Bloch, I., *Odoratus Sexualis: A Scientific and Literary Study of Sexual Scents And Erotic Perfumes*, American Anthropological Society, 1933

Corbin, A., *The Foul and the Fragrant: Odour and the French Social Imagination*, HUP, 1986

Chen, S. W. C., Shemyakin, A. and Wiedenmayer, P., 'The Role of the Amygdala and Olfaction in Unconditioned Fear in Developing Rats', *The Journal of Neuroscience*, 26.1 (4 January 2006), 233–240

Chu, S. and Downes, J., 'Odour Evoked Autobiographical Memories: Psychological Investigations of Proustian Phenomena', *Chemical Senses*, 25 (2000), 111–116

Classen, C., Howes, D. and Synnott, A., *Aroma: The cultural history of smell*, Routledge, 1994

Cromie, W., 'Scientists Find Evidence for A Sixth Sense in Humans', *Harvard Gazette Archives*, (20 May 1999)

Dach, C., 'The Scent of Fear', *AdBusters*, September / October 2006

Dalton, P., 'Body Odors as Biomarkers for Stress', *The Journal of Credibility Assessment and Witness Psychology*, 7.2 (2006), 116–126

Doty, R. (ed.), *Handbook of Olfaction and Gustation*, 2nd edition, Marcel Dekker, 2003

Doty, R., Wudarski, T., Marshall, D. and Hastings, L., 'Marijuana Odor Perception: Studies Modelled From Probable Cause Cases', *Law and Human Behaviour*, 28.2 (April 2004), 223–233

Freedman, A., 'Search is on for Emotion-Eliciting Scents', *Wall Street Journal*, 13 October 1988, section B

Gonzalez-Crussi, F., *The Five Senses*, Picador, 1989

Greenwood, J., *Curiosities of Savage Life*, London, 1864

Gilbert, A. N., *Compendium of Olfactory Research: explorations in aroma-chology : investigating the sense of smell and human response to odors, 1982–1994*, Kendall/Hunt Pub. Co., 1995

Graves, R., 'Total loss of smell occasioned by exposure to a very strong and disagreeable odour', *Dublin Journal of Medical and Chemical Science*, 6 (1834), 60

Graves, R., 'On the Treatment of Various Diseases', *Dublin Journal of Medical and Chemical Science*, 6 (1835) 60

Jaeger, G., *Entdeckung der Seele*, 2 vols, 1884–1885

Kaye, J. N., 'Symbolic Olfactory Display', MIT 2001, http:// alumni. media.mit.edu/~jofish/thesis/symbolic_olfactory_display (accessed on 18 December 2007)

Kenneth, J. H., Osmics: *The Science of Smell*, Oliver & Boyd, 1922

Kleiner, K., 'Sweet Smell of Success: How to beat the bacteria that cause body odour', *New Scientist*, 25 May 2002

Kohl, J., Atzmueller, M., Fink, B. and Grammer, K., 'Human Pheromones: Integrating Neuroendocrinology and Ethology', *Neuroendocrinology Letters*, 22.5 (2001), 309–321

Laird, D. A., 'Man's individuality in odor', *Journal of Abnormal and Social Psychology*, 29 (1934–35)

Lalumiere, M. L. and Quinsey, V. L., 'Pavlovian conditioning of sexual interests in human males', *Archives of Sexual Behaviour*, 27.3 (1998), 241–52

Le Guérer, A., *Scent: The mysterious and essential powers of smell*, translated by R. Miller, Chatto, 1992

Lengyel, O., Five Chimneys: *The Story of Auschwitz*, Siff-Davis Publishing, 1947

McCartney, W., *Olfaction and Odours*, Springer-Verlag, 1968

Maccord, C. and Wetheridge, W., Odors: *Physiology and Control*, McGraw-Hill Book Co., 1949

McKenzie, D., *Aromatics and the soul: a study of smells*, William Heinemann, 1923

Monin, E., *Essai sur les Odeurs Du Corps Humain*, Georges Carré, 1885

Motlok, A., 'The Scent of Fear', *New Scientist*, 1 May 1999

Piesse, C. H., *Olfactics and the Physical Senses*, Piesse & Lubin, 1887

Rasch, B., Büchel, G., Born, J., 'Odor Cues During Slow-Wave Sleep Prompt Declarative Memory Consolidation', *Science*, 315 (9 March 2007)

Rindisbacher, H. J., *The Smell of Books: a cultural-historical study of olfactory perception in literature*, The University of Michigan Press, 1992

Rogowski, M., 'An Attempt to Determine the Possibility of Transfer of a Person's Scent Onto A Carrier Through the Medium of His or Her Garment Used by Another Person and Infiltration of Scent through the Garment', *Problems of Forensic Sciences*, 50 (2002) 64–77

Schoon A. and Haak R., K9 *Suspect Discrimination: Training and Practicing Scent Identification Line-Ups*, Detselig Enterprises Ltd, 2002

Sidoli, M., 'Farting as a defence against unspeakable dread', *Journal of Analytical Psychology*, 41.2 (April 1996), 165–178

Smith, K., Van Toller, S. and Dodd, G.H., 'Unconscious odor conditioning in human subjects', *Biological Psychology*, 17 (1983), 221

Stoddart, M.D., *The Scented Ape: The Biology and Culture of Human Odour*, CUP, 1990

Sullivan, R., 'Developing a Sense of Safety: The Neurobiology of Neonatal Attachment', *Annals of the New York Academy of Sciences*, 1008 (2003), 122–131

Takahashi, K.L., Nakashima, B.R., Hong, H. and Watanabe, K., 'The smell of danger: A behavioural and neural analysis of predator odor-induced fear', *Neuroscience and Biobehavioural Reviews*, 29 (2005), 1157–1167

Thomas, L., *The Lives of a Cell: Notes of a Biology Watcher*, Viking, 1974

Thomas, L., 'On Smell', *Late Night Thoughts*, OUP, 1984

Van Toller, S. and Dodd, G. H. (eds), *Perfumery, the psychology and biology of fragrance*, Chapman and Hall Ltd, 1988

Watson, L., *Jacobson's organ and the remarkable nature of smell*, W.W. Norton, 2000

Wojcikiewicz, J., 'Dog Scent Lineup as Scientific Evidence', *Problems of Forensic Sciences*, 41 (2000)

Zelano, C. and Sobel, N., 'Humans as an Animal Model for Systems-Level Organization of Olfaction', *Neuron*, 48 (2005), 431–454

Bees

Bishop, H., *Robbing the Bees*, Simon and Schuster, 2005

Galizia, C. G., 'Brainwashing, Honeybee Style', Science 317.5836 (20 July 2007), 326–237

Robinson, G. E., 'Chemical Communication in Honeybees', *Science*, 271.5257 (29 March 1996), 1824–1825

Vergoz, V., Schreurs, H. A., Mercer, A. R., 'Queen Pheromone Blocks Aversive Learning in Young Worker Bees', *Science*, 317.5836 (20 July 2007), 384–386

Von Frisch, K., 'Decoding the Language of the Bee', Nobel Lecture, 12 December 1973

Is Beekeeping for You? http://www.beemaster.com/honeybee/youbees.html (accessed on 18 December 2007)

Wadey, H.J., *The Behaviour of Bees and of Bee-keepers*, W. & J. Mackay & Co., 1948

Dogs

ACPO Police Dog Working Group (2002) *Police Dog Training and Care Manual*: http://www.acpo.police.uk/asp/policies/policieslist.asp (accessed on 28 March 2006)

Arundel, R., 'The Police Dog: His Selection and Training', *Police Journal*, 1 (1928)

Close Quarter Battle K-9™, 'Counter-Terrorist K-9': http://www.cqbk9.com/counter_terrorist_k9.html (accessed on 19 December 2007)

Cotter, M., *All About German Shepherd Dogs*, Castle Warden, 2000

Davidson, O.G., 'The Secret file of Abu Ghraib', *Rolling Stone*, 28 July 2004

Fogle, B., *The Dog's Mind*, Pelham Books, 1964

Gersbach, R., *Manuel de Dressage des Chiens de Police*, translated by Elmer, D., Fournier, undated

Grainger, D., 'Sit! Stay! Testify!', *Fortune*, 26 January 2004

Helm, S., 'The Nazi Guard's Untold Love Story', *Sunday Times*, 5 August 2007

Hart, L.A., Zasloff R.L., Bryson, S., Christensen, S.L., 'The role of police dogs as companions and working partners', *Psychological Reports*, 86.1 (February 2000), 190–202

Kerr, I. and McGill, J., 'Emanations, Snoop Dogs and Reasonable Expectations of Privacy', *Criminal Law Quarterly*, 52.3 (2007)

Lemish, M., *War Dogs: Canines in Combat*, Brassey's, 1996

Lubow, R.E., *The War Animals*, Doubleday & Company, 1977

Marks, A., 'Drug Detection Dogs and the Growth of Olfactory Surveillance: Beyond the Rule of Law?', *Surveillance & Society*, 4.3 (2007), 257–271

Masson, J.F., *Dogs Never Lie about Love*, Johnathan Cape, 1997

Maynard Leonard, R. (ed.), *The Dog in British Poetry*, David Nutt, 1893

Nash, M., 'Who Let The Dogs In?', *Journal of Psychiatric and Mental Health Nursing*, Vol 12 (6), (December 2005), 745

New South Wales Ombudsman, *Discussion Paper: Review of the Police Powers (Drug Detection Dogs) Act*, NSW Government Publication, (June 2004)

New South Wales Ombudsman, *Review of the Police Powers (Drug Detection Dogs) Act 2001*, NSW Government Publication (September 2006)

Orr-Munro, T., 'Dogs of Law', *Police Review*, (8 June 2003)

Prison Service News, 'It's a Dog's Life', *Prison Service News*, (October 2001)

Richardson, E.H., *British War Dogs: Their Training and Psychology*, Skeffington & Son, undated

Richardson, E.H., *War, Police, and Watch Dogs*, William Blackwood and Sons, 1910

Sheldrake, R., *Dogs That Know When Their Owners Are Coming Home*, Arrow Books, 1999

Stark James, 'On the influence of colour on odours', *Dublin Journal of Medical and Chemical Science*, 6 (1835), 127–134

Stubbs, C., 'Hell Hounds: In Search of Britain's Phantom Black Dogs', *Fortean Times*, (April 2005)

Other Animals and Plants

Backster, C., 'Evidence of a Primary Perception in Plant Life', *International Journal of Parapsychology*, 10. 4 (Winter 1968), 329–348

Carwardine, M., *Extreme Nature*, HarperCollins, 2005

Evans, E.P., *The Criminal Prosecution and Capital Punishment of Animals: The Lost History of Europe's Animal Trials*, Faber and Faber, 1988

Hughes, H.C., *Sensory Exotica: A world beyond human experience*, MIT Press, 2001

Pomery, C., *State Secrets: Behind the Scenes of the 20th Century*, National Archives, 2006

Sullivan, R., *Rats: A Year with New York's Most Unwanted Inhabitants*, Granta Books, 2004

Taylor, D., *David Taylor's Animals in Battle*, Boxtree, 1989

Mind and Brain

Chomsky, N., 'The Case Against B.F. Skinner', *The New York Review of Books*, (30 Dec. 1971)

Flanagan, O., *The Science of the Mind*, The MIT Press, 2nd edition, 1991

Gibb, B. J., *The Rough Guide to the Brain*, Rough Guides Ltd, 2007

Gregory, R.L. (ed.), *The Oxford Companion to the Mind*, OUP, 2004

Hunter, E., *Brainwashing: From Pavlov to Powers*, The Bookmailer, 1965

Koestler, A., *The Ghost in the Machine*, Hutchinson & Co., 1967

LeDoux, J., *The Emotional Brain*, Phoenix, 1998

Moreno, J.D., *Mind Wars: Brain Research and National Defense*, Dana Press, 2005

Narby, J., *Intelligence in Nature*, Penguin, 2005

Pavlov, I.P., *Conditioned Reflexes: An investigation of the physiological activity of the cerebral cortex*, Humphrey Milford, 1927

Taylor, K., *Brainwashing: The Science of Thought Control*, OUP, 2004

Surveillance, Law and Chemical and Biological Weapons

Albert, S. and Hitt W., *Intercultural Differences in Olfaction, Remote Area Conflict Information Center*, Battelle Memorial Institute, 1966

British Medical Association, *The use of drugs as weapons: the concerns and responsibilities of healthcare professionals*, British Medical Association, 2007

Brownsword, R. and Yeung, K. (eds), *Regulating Technologies*, Hart Publishing, 2008 forthcoming

Cousens, M., Surveillance Law, Lexis Nexis Butterworths, 2004

Dando, M., *A New Form of Warfare: The Rise of Non-Lethal Weapons*, Brassey's 1996

Davison, N., *'Off the Rocker' and 'On the Floor': The Continued Development of Biochemical Incapacitating Weapons*, Bradford Science and Technology Report 8, Aug. 2007

Denning, A, *The Road to Justice*, Stevens & Sons, 1955

Department of Trade and Industry, *Turning the Corner: Final Report of the Foresight Crime Panel*, DTI, 2001

Funder, A., *Stasiland*, Granta Books, 2003

Ekblom, P., 'Can we make crime prevention adaptive by learning from other evolutionary studies?', *Studies in Criminology and Crime Prevention*, 8 (1999), 27–51

Ebisike, N., *An Appraisal of Forensic Science Evidence in Criminal Proceedings*, Greenway Publishers, 2005

Farrell, G., 'Skunks, Cinnabar Moths and Smart Policing', *Police and Government Security Technology Journal*, (1997), 62–63

Greenberg, K.J. and Dratel, J.L., *The Torture Papers: The Road to Abu Ghraib*, CUP, 2005

Hayes, B., *Arming Big Brother: The EU's Security Research Programme*, Transnational Institute, The Briefing Series 1 (2006)

House of Commons Select Committee on Science and Technology, *Forensic Science on Trial*, Seventh Report of Session 2004–2005, The Stationery Office Ltd.

Koplow, D., *Non-Lethal-Weapons*, CUP, 2006

Langley, C., *Soldiers in the Laboratory*, Scientists for Global Responsibility, 2005

Lyon, D. (ed.), *Theorizing Surveillance: The Panoptican and Beyond*, Willan Publishing, 2006

Marks, J.D., *The Search for the Manchurian Candidate*, Norton, 1991

Marx, G.T., 'What's new about the "new surveillance"? Classifying for change and continuity', *Surveillance & Society 1.1* (2002), 9–29.

Minton, A., *The Privatisation of Public Space*, Royal Institution of Chartered Surveyors, 2006

O'Harrow, R., *No Place to Hide*, Penguin Books, 2006

Pease, K. 'Science in the Service of Crime Reduction', *Handbook of Crime Prevention and Community Safety*, N.Tilley (ed.), Willan Publishing, 2005

Rappert, B., *Non-Lethal Weapons as Legitimizing Forces? Technology,*

Politics and the Management of Conflict, Frank Cass, 2003

Rischikof, H. and Schrage, M., 'Technology vs Torture', http://www.slate.com/id/2105332/, accessed on 15 October 2007

Ronson, J., *The Men Who Stare At Goats*, Picador, 2004

Royal Society of Chemistry, Report from the *Chemical Science and Crime Prevention Workshop*, 2004

Stafford-Smith, C., *Bad Men: Guantanamo Bay and the Secret Prisons*, Weidenfeld & Nicolson, 2007

Sunshine Project, *Non-lethal weapons research in the US: Calmatives and Malodorants* (July 2001)

US Army and Monell Chemical Senses Center, *Establishing Odor Response Profiles*, Contract DDAD13–98-M0064 (April 1998)

Wheelis, M. and Dando, M., 'Neurobiology: A case study of the imminent militarization of biology', *International Review of the Red Cross*, 87.859 (September 2005)

Fiction

Atwood, M., *Oryx and Crake*, Anchor Books, 2003

Ballard, J.G., *Super-Cannes*, Harper Perennial, 2006

Byatt, A.S., *Angels & Insects*, Vintage, 1995

Wilkie Collins, W., *The Haunted Hotel: A Mystery of Modern Venice*, Chatto & Windus, 1879

Dahl, R., *Switch Bitch*, Penguin Books, 1976

Doyle, A.C., *The Hound of the Baskervilles*, George Newnes, 1902

Gloag, J., *The New Pleasure*, George Allen & Unwin, 1933

Herbert, F., *Hellstrom's Hive*, Bantam Books, 1972

Huxley, A., *Brave New World*, Chatto & Windus, 1932

Huxley, A., *Ape and Essence*, Chatto & Windus, 1949

Kesey, K., *One Flew Over the Cuckoo's Nest*, Picador, 1962

Koestler, A., *Arrival and Departure*, Macmillan, 1943

Orwell, G., *1984*, Secker & Warburg, 1949

Robbins, T., *Jitterbug Perfume*, Bantam Books, 1991

Süskind, P., *Perfume*, Penguin Books, 1985

Thompson, H.S., *Fear and Loathing in Las Vegas*, Straight Arrow Books / Quick Fox, 1973

Wells, H.G., *The Island of Dr Moreau*, Ernest Benn, 1927

Miscellaneous

Ball, P., *Stories of the Invisible: a guided tour of molecules*, OUP, 2001

Boring, E., 'Psychology of Perception; importance in war effort', *American Journal of Psychology*, 55 (1942), 423

Chomsky, N., 'Resort to Fear', Foreword to Nirmalangushu Mukheriji's *December 13: Terror Over Democracy*, New Delhi: Promilla and Co. Publications and Bibliophile South Asia, 2005

Diechmann, U., *Biologists under Hitler*, translated by Thomas Dunlop, HUP, 1966

Frenay, R., Pulse: *The coming age of systems and machines inspired by living things*, FarrarStraus and Giroux, 2006

Fromm, E., *Fear of Freedom*, Routledge, 1942

Horowitz, M. and Palmer, C. (eds), *MOKSHA, Aldous Huxley's Classic Writings on Psychedelics and the Visionary Experience*, Park Street Press, 1977

Huxley, A. (ed.), *Encyclopaedia of Pacifism*, Chatto & Windus, 1937

Huxley, A., *Brave New World Revisited*, Chatto & Windus, 1959

Huxley, A., *Science and Literature*, Chatto & Windus, 1963

Immelmann, K. and Beer, C., *A Dictionary of Ethology*, HUP, 1992

Koestler, A., *The Case of the Midwife Toad*, Pan Books, 1971

Pieper, W., *Mark Twain's Guide to Heidelberg*, Werner Pieper's MedienXperimente, 1995

Sullivan, J.W.N., *Limitations of Science*, Penguin Books, 1933

Welsford, E., *The Fool: His Social and Literary History*, Faber and Faber, 1935

Internet Resources

Flex Your Rights Foundation homepage: http://www.flexyourrights.org

Information Commissioner's Office: http://www.ico.gov.uk

NO2ID homepage: http://www.no2id.net/index.php

Nuffield Council on Bioethics: http://www.nuffieldbioethics.org

Sense of Smell Institute: http://www.senseofsmell.org/resources/acr_toc.php

Sunshine Project homepage: http://www.sunshineproject.org

Surveillance & Society Homepage: http://www.surveillanceand-society.org

Spy Blog homepage: SpyBlog.org.uk

University of Bradford Disarmament Research Centre's Project on Strengthening the Biological and Toxin Weapons convention: http://www.brad.ac.uk/acad/ sbtwc

Acknowledgements

I'd like to thank Lizzy Kremer and Laura West at David Higham Associates, Ed Faulkner, Davina Russell and everyone at Virgin Books, Jon Banes, Mark Ballard, Nathan and Lisa Bateman, Glenn and Alicia Black, Tom Bowring, Toby and Laura Charkin, Caroline Coon, Ray D., Hywell Dinsdale, Dick Doty, David Furlow, Monica Garrido, Matthias Hauke, Joe Jacob, Dieter Just, Katrin Kamolz, Marty Langford, Patrick, Howard, Francesca and Judy Marks, Milly Marston, Tracey and Jonathan Moberly, Richard Parry, Werner Pieper, Dave Pollard, Cassandra Purdy, Rachel Reasbeck, Andrew Robertson, Chris Roe, Anabel Rosello, John Smith, John and Veena, Franck Pham-Van, Jacob and Izaac Sanderswood, Oona and all those who helped me who wish to remain anonymous.

Thank you to the following for permission to quote material:

Oxford University Press, for the excerpt from Katherine Taylor's *Brainwashing*; Bloomsbury for the excerpt from Margaret Atwood's *Oryx and Crake*; The Economist Newspaper Limited; Lyall Watson and W. W. Norton & Company; *Nineteen Eighty Four* by George Orwell (Copyright © George Orwell, 1949) by permission of Bill Hamilton as the Literary Executor of the Estate of the Late Sonia Brownell Orwell and Secker & Warburg Ltd; Roald Dahl, Jonathan Cape Ltd & Penguin Books Ltd.